Lecture Notes in Business Information Processing

254

More information about this series at http://www.springer.com/series/7911

Raúl León · María Jesús Muñoz-Torres
Jose M. Moneva (Eds.)

Modeling and Simulation in Engineering, Economics and Management

International Conference, MS 2016
Teruel, Spain, July 4–5, 2016
Proceedings

 Springer

Editors
Raúl León
Department of Accounting and Finance
University of Zaragoza
Zaragoza
Spain

Jose M. Moneva
Department of Accounting and Finance
University of Zaragoza
Zaragoza
Spain

María Jesús Muñoz-Torres
University Jaume I
Castellón de la Plana
Spain

ISSN 1865-1348   ISSN 1865-1356   (electronic)
Lecture Notes in Business Information Processing
ISBN 978-3-319-40505-6   ISBN 978-3-319-40506-3   (eBook)
DOI 10.1007/978-3-319-40506-3

Library of Congress Control Number: 2016941607

Printed on acid-free paper

This Springer imprint is published by Springer Nature
The registered company is Springer International Publishing AG Switzerland

Preface

The Association for the Advancement of Modelling and Simulation Techniques in Enterprises (AMSE) and the University of Zaragoza are pleased to present the main results of the International Conference on Modelling and Simulation in Engineering, Economics, and Management, held in Teruel, Spain, July 4–5, 2016, through this book of proceedings published with Springer in the series *Lecture Notes in Business Information Processing*.

MS 2016 Teruel was co-organized by the AMSE Association and the University of Zaragoza through the GESES Research Group, with the support of the SoGReS-MF Research Group from University Jaume I. It offered a unique opportunity for students, researchers, and practitioners to present and exchange ideas concerning modelling, simulation, and related topics, and see how they can be implemented in the real world.

In this edition of the MS international conference, we paid special attention to the area of modelling and simulation in business and economic research. The title of the book, *Modelling and Simulation in Engineering, Economics and Management,* refers to a broad research area and we have tried to systematize the resulting collection into a reasonable number of cohesive themes. The importance of modelling and simulation techniques in engineering, business, and economic research is reflected through several contributions related to modelling and simulation in industrial development, finance, accounting, business management, marketing, economics, and politics, among others. This congress has as main incentive the interaction among different disciplines and this interaction is helping to improve the availability of models and techniques in business and economic research.

The MS 2016 Teruel proceedings comprise 20 papers—with a total of 56 authors from nine countries—selected from 52 submissions. The book is organized according to three general tracks of the congress: Modelling and Simulation in Finance and Accountability, Modelling and Simulation in Business Management and Economics, and Modelling and Simulation in Engineering and Other General Applications.

We would like to thank all the contributors, reviewers, plenary speakers, and the scientific and honorary committees for their kind co-operation with MS'2016 Teruel; the Organizing Committee for their support regarding the logistics of the conference; and Ralf Gerstner, Alfred Hofmann, Christine Reis, and Eléonore Samklu (Springer) for their kind advice and help in publishing this volume.

Finally, we would like to express our gratitude for their support in the preparation of this book to Springer and in particular to the editors of the book series *Lecture Notes in Business Information Processing,* Wil van der Aalst, John Mylopoulos, Michael Rosemann, Michael J. Shaw, and Clemens Szyperski.

July 2016

Raúl León
María Jesús Muñoz-Torres
Jose M. Moneva

Preface

Organization

Honorary Committee

Jaime Gil Aluja   President AMSE, Spain
Lotfi Zadeh   University of Berkeley, USA

Organizing Committee

Anna M. Gil Lafuente, Spain, Chair
Jose M. Moneva, Spain, Chair
María Jesús Muñoz Torres, Spain, Chair
Raúl Leon Soriano, Spain, Chair
César Leonardo Guerrero Luchtenberg, Spain
Cristina Ferrer, Spain
Eduardo Ortas, Spain
Encarna Esteban, Spain
Francisco Javier Arroyo Cañada, Spain
Inmaculada Plaza, Spain
José M. Merigó Lindahl, Chile
Juan Pablo Maicas, Spain
Juana María Rivera Lirio, Spain
María Ángeles Fernández Izquierdo, Spain

Scientific Committee

Angel A. Juan, Spain
Anna M. Gil-Lafuente, Spain
Antonio Terceño Gómez, Spain
Beatriz Flores Romero, Mexico
Carlos Tomás Medrano Sánchez, Spain
César Leonardo Guerrero Luchtenberg, Spain
Christian Berger-Vachon, France
Eduardo Ortas Fredes, Spain
Egon Zizmon, Slovenia
Elena Escrig Olmedo, Spain
Emilio Vizuete Luciano, Spain
Federico González Santoyo, Mexico
Francesco Carlo Morabito, Italy
Gerardo Gabriel Alfaro Calderón, Mexico
Idoya Ferrero Ferrero, Spain

Sponsors

Association for the Advancement of Modelling
& Simulation Techniques in Enterprises

1542

UNIVERSIDAD DE ZARAGOZA
Grupo de Estudios Sociales y
Económicos del Tercer Sector

UNIVERSITAT
JAUME·I

Sustainability of Organizations
and Social Responsibility Management-
Financial Markets

Fundación
Universitaria
Antonio Gargallo

Vicerrectorado para
el Campus de Teruel
Universidad Zaragoza

Contents

Engineering and Other General Applications

Modeling and Simulation in Finance and Accounting

A SimILS-Based Methodology for a Portfolio Optimization Problem with Stochastic Returns

Laura Calvet[1]([⊠]), Renatas Kizys[2],
Angel A. Juan[1], and Jesica de Armas[1]

[1] Computer Science Department, Open University of Catalonia-IN3,
Barcelona, Spain
{lcalvetl,ajuanp,jde_armasa}@uoc.edu
[2] Subject Group of Economics and Finance, University of Portsmouth,
Portsmouth, UK
renatas.kizys@port.ac.uk

Abstract. Combinatorial optimization has been a workhorse of financial and risk management, and it has spawned a large number of real-life applications. Prominent in this body of research is the mean-variance efficient frontier (MVEF) that emanates from the portfolio optimization problem (POP), pioneered by Harry Markowitz. A textbook version of POP minimizes risk for a given expected return on a portfolio of assets by setting the proportions of those assets. Most authors deal with the variability of returns by employing expected values. In contrast, we propose a simILS-based methodology (i.e., one extending the Iterated Local Search metaheuristic by integrating simulation), in which returns are modeled as random variables following specific probability distributions. Underlying simILS is the notion that the best solution for a scenario with expected values may have poor performance in a dynamic world.

Keywords: Portfolio optimization · SimILS · Metaheuristics · Simulation

1 Introduction

Investments play an essential role in our society through wealth creation, sustainable economic growth and ultimately improvements in welfare standards. They provide companies with the necessary funds to transform ideas and resources into profitable projects, social benefits and jobs. From the point of view of a portfolio investor, POP is a strategy of (a) selection of financial assets and (b) determination the optimal weights allocated to those assets that results in a desired portfolio return and an associated minimum level of risk. This combinatorial optimization problem (COP) is known as the portfolio optimization problem (POP), a milestone of modern portfolio theory, founded by Markowitz [1]. Key to POP is a quadratic objective function that is (a) computed by aggregating over the covariances of the constituent asset returns, and (b) minimized subject to a desired rate of return. It is worth noting that other risk measures have been applied in the literature such as value-at-risk. Additionally, portfolio weights must add up to one and, in most cases, take on non-negative values. A realistic POP introduces further constraints. In particular, pre-assignment, quantity

© Springer International Publishing Switzerland 2016
R. León et al. (Eds.): MS 2016, LNBIP 254, pp. 3–11, 2016.
DOI: 10.1007/978-3-319-40506-3_1

and cardinality constraints have received overwhelming attention in extant literature. The pre-assignment constraint allows the investor to pre-select some assets, irrespective of their risk-return characteristics. The quantity constraint confines the weight allocated to an asset in the portfolio within a desired range of values. One the one hand, the upper limit (the ceiling) of the range attempts to reduce the exposure to each asset. On the other hand, the lower limit (the floor) rules out investments in negligible quantities, which may be prohibitively costly, since the transaction costs may reduce or erase the benefit. While recognising that this constraint arises as a result of the investor's discretionary decisions, it has received growing interest. For instance, [2] argue that its inclusion can lead to improved out-of-sample performance of optimization performance, can help contain portfolio volatility, boost realized portfolio performance, as well as decrease downside risk and shortfall probability. Finally, the cardinality constraint sets a minimum and maximum value for the number of assets in the portfolio. The lower bound aims to diversify the investment, i.e., allocate resources to a set of imperfectly correlated assets. Such strategy seeks to minimize the overall risk of portfolio investment. The upper bound is dictated by the evidence that marginal benefits of portfolio diversification starts to decrease after the number of assets already selected in the portfolio hits a certain threshold [3]. In addition, portfolios with a large number of assets are more costly in terms of complexity, managerial effort and the ensuing increased transaction costs. These constraints make the problem NP-hard [4].

Optimization methods may be classified into exact methods and heuristics/ metaheuristics [5]. The first group includes procedures that guarantee the optimality of a solution. However, exact methods may require making strong assumptions or large amounts of time, especially when they are used to solve real-life complex problems. Within the second group, heuristics are experience-based procedures, which usually provide near-optimal solutions in considerably less time. By contrast, metaheuristics [6] are general templates, which may solve a broad range of problems without having to be tailor-made to a particular problem and often in real time. In the literature on the portfolio optimization, linear [7] and quadratic [8] programming methods have been predominant exact methods. However, due to the complexity of these problems, metaheuristics are increasingly more employed at present [9].

Despite the non-exhaustive nature of applications of realistic POP, they have not been extensively studied. As aforementioned, a textbook version of POP underlies the empirically unsupported assumption of constant expected rate of return, a key limitation in a large and growing body of research. The main contribution of this research is to address this limitation. Indeed, since asset return is a random variable that obeys a certain probability density function, and future returns are unpredictable, the minimum desired rate of return may not be attained with certainty. More concretely, we relax the above simplifying assumption and randomize the minimum desired rate of return. The resulting problem is referred to as the *Stochastic* POP (SPOP). The solver that is constructed to solve SPOP is relatively new and is called SimILS [10]. It envisages an extension of the Iterated Local Search (ILS) metaheuristic [11] that integrates simulation techniques to address sources of uncertainty embedded in a randomized objective function or/and budget constraint. In short, while a metaheuristic searches for high-quality solutions for a deterministic version of the problem, which employs expected values of random variables, simulation techniques are applied to test them in a stochastic environment. In fact, this approach – coined simheuristics [12] – suggests combining metaheuristics and simulation techniques. In this context, our research aims to: *(i)* derive a mathematical formulation for the Stochastic POP, *(ii)* develop a solving methodology based on an existing algorithm for the POP [13]; and *(iii)* illustrate its use by solving a benchmark instance.

This paper is organized as follows. Section 2 provides a formal description of the problem. Section 3 proposes a methodology. A computational experiment is carried out in Sect. 4, while the results are analyzed in Sect. 5. Finally, Sect. 6 gathers the main conclusions.

2 Description of the Problem

Let there be a set $A = \{a_1, a_2, \ldots, a_n\}$ of n assets, where each asset a_i ($\forall i \in \{1, 2, \ldots, n\}$) is characterized by an expected return r_i. The covariance between two assets a_i and a_j ($\forall i, j \in \{1, 2, \ldots, n\}$) is denoted by σ_{ij}. A solution for the POP (see Fig. 1), is a vector $X = (x_1, x_2, \ldots, x_n)$, where each element x_i ($0 \leq x_i \leq 1$) represents the weight of the asset a_i in the portfolio. The aim of the POP is to minimize the risk of the investment and obtain an expected return greater than a specific threshold R. A realistic version includes also the pre-selection, quantity and cardinality constraints. The pre-selection constraint dictates whether an asset a_i must be in the solution (i.e., $x_i > 0$) by means of the parameter p_i: $p_i = 1$ if a_i is pre-selected, and $p_i = 0$ otherwise. The quantity constraint specifies for each asset a_i a lower and an upper bound, ε_i and δ_i ($0 \leq \varepsilon_i \leq \delta_i \leq 1$), respectively. The cardinality constraint sets the lower and upper limits on the number of assets included in the portfolio, k_{min} and k_{max} ($1 \leq k_{min} \leq k_{max} \leq n$), respectively. One key difference between the stochastic and classical versions of POP is that the former assumes uncertain future returns on a portfolio of assets, with a certain probability of not attaining the threshold value.

$$\min f(X) = \sum_{i=1}^{n} \sum_{j=1}^{n} \sigma_{ij} x_i x_j$$

Subject to :

$$P\left(\sum_{i=1}^{n} R_i x_i \geq R\right) \geq P_0 \tag{1}$$

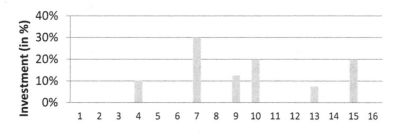

Fig. 1. Example of a solution representation

$$\sum_{i=1}^{n} x_i = 1 \qquad (2)$$

$$\varepsilon_i z_i \leq x_i \leq \delta_i z_i, \quad \forall i \in \{1, 2, \ldots, n\} \qquad (3)$$

$$0 \leq \varepsilon_i \leq \delta_i \leq 1, \quad \forall i \in \{1, 2, \ldots, n\} \qquad (4)$$

$$z_i \leq M x_i, \quad \forall i \in \{1, 2, \ldots, n\} \qquad (5)$$

$$p_i \leq z_i, \quad \forall i \in \{1, 2, \ldots, n\} \qquad (6)$$

$$k_{min} \leq \sum_{i=1}^{n} z_i \leq k_{max} \qquad (7)$$

$$z_i \in \{0, 1\}, \quad \forall i \in \{1, 2, \ldots, n\} \qquad (8)$$

The objective function minimizes the risk of the investment. Equation (1) guarantees that the return on investment will be no smaller than the threshold R with a probability of at least P_0. Equation (2) restrains portfolio investment to the existing resources. An auxiliary variable is introduced to indicate whether the asset a_i is in the solution ($z_i = 1$ in this case, $z_i = 0$ otherwise). For each asset a_i, Eq. (3) sets a lower and an upper bound (ε_i and δ_i, respectively) for x_i, in case the asset is selected (i.e., $z_i = 1$). The two bounds range from 0 and 1 (Eq. 4). In Eq. (5) M is a very large positive value such that $M x_i \geq 1$ for all i if $x_i > 0$. Equation (6) defines the pre-assignment constraint, where z_i depends on the parameter p_i. If the asset a_i is pre-selected (i.e., $p_i = 1$), then it also appears in the solution (i.e., $z_i = 1$). Equation (7) describes the cardinality constraint. Finally, Eq. (8) defines z_i as a binary variable.

3   Our Methodology

The proposed methodology follows a simILS approach. It is demonstrated to be successful for solving realistic COPs with sources of uncertainty [14]. It is a natural extension of ILS-based algorithms to address stochastic COPs. Specifically, we combine a solving methodology for the POP [13] with Monte Carlo simulation (MCS) techniques. The referred work describes a powerful yet simple algorithm, which includes heuristics for the selection of assets and a quadratic programming solver that allocates weights to POP. It uses memory caches to enhance the algorithm's performance. In fact, it provides high-quality solutions in real time, only within a few seconds.

Our methodology is summarized in Fig. 2 and described next.

First, the stochastic instance is transformed into deterministic by replacing random variables by their means. Second, an initial solution by means of the algorithm described in [13]. It constructs a solution by combining the pre-selected assets with high-return assets. The solution has to be feasible in the stochastic environment, i.e., the required return (R) has to be reached with a probability no smaller than P_0. MCS is employed to estimate this probability by means of the proportion of cases in a sample of generated scenarios where the return obtained is at least as high as R. Each scenario

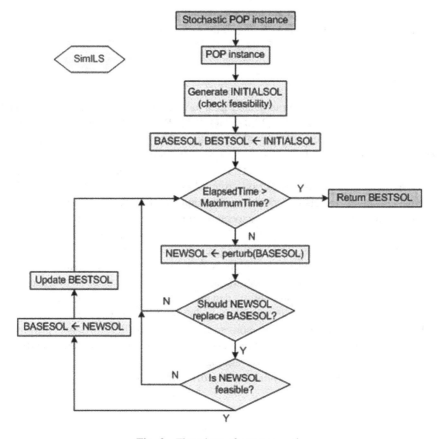

Fig. 2. Flowchart of our approach

is created by randomly drawing a value for each return in the original instance. Third, copies of the initial solution are stored as base and best solutions. Fourth, of the above specified steps are repeated until a stopping criterion based on the elapsed time is met. First, a new solution is created by 'perturbing' the base solution. This perturbation, defined in the original algorithm, randomly erases some assets from the portfolio and introduces others. An acceptance criterion is introduced to determine whether the new solution is promising and should replace the base solution or should be discarded. It is a Demon-like acceptance criterion [5], which accepts the solutions that improve the objective function value (i.e., the risk) and those that worsen it but satisfy the following conditions; (*i*) no consecutive deteriorations take place, and (*ii*) the degradation does not exceed the value of the last improvement. The next step consists of checking the feasibility of the new solution as before, using MCS. Only if it is feasible, the new solution is copied into the base solution, and the best solution is updated (if improved). Finally, the best solution is returned.

Note that this approach presents relevant advantages: (*i*) its modularity, which enables the reuse of problem-specific procedures from the original algorithm, (*ii*) it is relatively simple to understand and implement, and (*iii*) it does not add too much time, since MCS is only used to check the feasibility of promising solutions.

4   Computational Experiments

Our methodology has been implemented as a Java application. A standard personal computer, Intel Core i5 CPU at 3.2 GHz and 4 GB RAM with Windows 7 has been employed. We have experimented with a stock market database from the repository ORlib (http://people.brunel.ac.uk/~mastjjb/jeb/orlib/portinfo.html), which was proposed in [15]. It represents the market index Hang Seng (Hong Kong) measured at weekly frequency spanning the period from March 1992 to September 1997. This benchmark instance gathers expected returns r_i and standard deviations σ_i. In order to assess our simheuristic methodology, expected returns have been replaced with random variables R_i that distribute normally with mean r_i and standard deviation σ_i. The instance contains 31 assets. The expected returns average is 0.0035 (95 % IC: 0.0027-0.0043) and the standard deviation is 0.002. The return standard deviations average 0.0457 (95 % IC: 0.0430–0.0484) and deviate from the mean on average by 0.0073. Figure 3 displays the probability density functions for some selected assets. Table 1 proves that there is a positive association between expected returns and standard deviation (Pearson product-moment correlation coefficient: 0.4973). In other words, investors expect a higher return for assets characterized by a higher risk. Return correlations among different assets average 0.5266 (95 % IC: 0.5137–0.5395), which suggests the presence of gains from portfolio diversification.

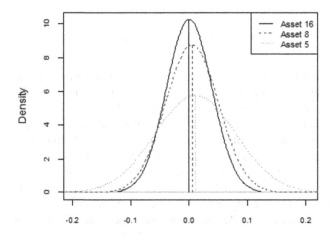

Fig. 3.   Returns of selected assets following Normal distributions (Color figure online)

Our algorithm is executed 10 times using different seeds; only the best results are shown. To minimize the computational time, the number of runs for assessing promising solutions has been set to 2000. The other parameters, including the time of the iterative procedure, have been set to the values suggested in [13].

Table 1. Correlation analysis between expected returns and standard deviation

Correlation coefficient	P-value	IC (95 %)
0.4973	0.0044	0.1736–0.7241

100 equidistant values for the required rate of return have been selected. Table 2 shows the first and the last 5 observations on the required return, risk and reliability (or probability of the return being no smaller than the required return) associated to the solution obtained by the original algorithm (i.e., considering expected values), the risk found with our methodology considering the probabilities of 0.48 and 0.52, and the gap between them. The solutions of our methodology were obtained in 4.783 s on average.

Table 2. Table of results

	Expected values		P_0: 0.48	P_0: 0.52	
Required return	Risk	Reliab. (%)	Risk (1)	Risk (2)	Gap (2)-(1) (%)
0.002861	0.000642	50.07	0.000642	0.000645	0.00024
0.002942	0.000643	50.28	0.000643	0.000645	0.00023
0.003023	0.000644	49.22	0.000644	0.000648	0.00047
0.003104	0.000644	49.55	0.000644	0.000646	0.00019
0.003185	0.000645	50.12	0.000645	0.000647	0.00011
0.010542	0.004194	49.02	0.004194	0.004194	0.00000
0.010622	0.004332	50.64	0.004332	0.004475	0.00014
0.010703	0.004475	49.90	0.004475	0.004475	0.00000
0.010784	0.004623	50.12	0.004623	0.004623	0.00000
0.010865	0.004776	49.67	0.004776	N/A	N/A

5 Analysis of Results

The computational results suggest that requiring returns above a given threshold with a higher probability leads to the same or higher portfolio variance. Moreover, the instance may become unsolvable.

Figure 4 shows the differences in terms of risk between the 0.48 and 0.52 probability solutions for each of the 100 equidistant values. Solutions associated to higher probability in general have greater risk. Figure 5 displays a multiple boxplot which depicts the distribution of returns obtained using MCS for a set of promising solutions when solving the instance for a specific required return. They are sorted according to the portfolio variance. Although no obvious differences can be identified, eyeballing suggests that the second solution have the smallest range. This case shows that it can be useful to provide a set of solutions to the decision-maker. Here, he would choose the first (risk: 0.0006444), which minimizes the risk, or the second (risk: 0.0006449), which has a slightly higher risk but with a lower return variability.

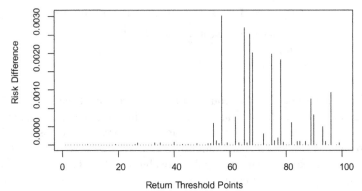

Fig. 4. Risk gaps considering two probabilities and 100 returns thresholds

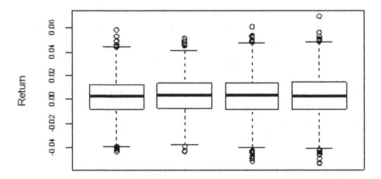

Fig. 5. Multiple Boxplot with returns distributions for several solutions

6 Conclusions

This work has addressed the Portfolio Optimization Problem (POP), which is a classic NP-hard Combinatorial Optimization Problem with plenty of applications. It consists in creating a portfolio selecting a subset of assets and setting their weights. Typically, authors solve this problem working with expected returns.

We have presented a mathematical formulation for the realistic POP, which considers the following constraints, commonly faced in real life: pre-assignment constraint (based on investor's preference), quantity constraint (which keeps each weight within user-specified floor and ceiling values) and cardinality constraint (providing a minimum and maximum value for the number of assets to include in the portfolio). Being a NP-hard problem, we require an approximate methodology for solving medium/high-sized instances in real time. Accordingly, we have proposed a simple methodology relying on the simILS approach. It combines an existing algorithm based on the Iterated Local Search metaheuristic for the classical version of the problem, which guides the search, with Monte Carlo simulation techniques, which checks the feasibility of promising solutions. A computational experiment employing an adapted benchmark instance is performed to illustrate its use and to analyze how the solutions change in terms of risk when varying the minimum required return and the probability of satisfying the constraint associated to this return.

Due to the stochasticity characterizing financial markets, the number of challenging versions of the POP, and their relevant applications, we plan to explore new methodologies or variants of the one presented to address problems in this field. For instance, an interesting line of research would be to consider more sources of stochasticity. Additionally, our methodology could be tested on instances describing different periods, countries, regions, sectors and asset classes.

Acknowledgments. This work has been partially supported by the Spanish Ministry of Economy and Competitiveness (TRA2013-48180-C3-P and TRA2015-71883-REDT), FEDER, and the Catalan Government (2014-CTP-00001).

References

1. Markowitz, H.M.: Portfolio selection. J. Finan. **7**, 77–91 (1952)
2. Kolm, P.N., Tütüncü, R., Fabozzi, F.J.: 60 years of portfolio optimization: practical challenges and current trends. Eur. J. Oper. Res. **234**, 356–371 (2014)
3. Maringer, D.: Portfolio Management with Heuristic Optimization. Springer, Heidelberg (2005)
4. Bienstock, D.: Computational study of a family of mixed-integer quadratic programming problems. Math. Program. **74**, 121–140 (1996)
5. Talbi, E.: Metaheuristics: From Design to Implementation. Wiley, New York (2009)
6. Boussaïd, I., Lepagnot, J., Siarry, P.: A survey on optimization metaheuristics. Inf. Sci. **237**, 82–117 (2013)
7. Mansini, R., Ogryczak, W., Speranza, M.G.: Twenty years of linear programming based portfolio optimization. Eur. J. Oper. Res. **234**, 518–535 (2014)
8. Sawik, B.: Bi-criteria portfolio optimization models with percentile and symmetric risk measures by mathematical programming. Przeglad Elektrotechniczny **88**, 176–180 (2012)
9. Adebiyi, A.A., Ayo, C.K.: Portfolio selection problem using generalized differential evolution 3. Appl. Math. Sci. **9**, 2069–2082 (2015)
10. Grasas, A., Juan, A., Lourenço, H.R.: SimILS: a simulation-based extension of the iterated local search metaheuristic for stochastic combinatorial optimization. J. Simul. (2015). doi:10.1057/jos.2014.25
11. Lourenço, H.R., Martin, O.C., Stützle, T.: Iterated local search. In: Glover, F., Kochenberger, G. (eds.) Handbook of Metaheuristics, pp. 321–353. Kluwer Academic Publishers, Norwell (2003)
12. Juan, A., Faulin, J., Grasman, S., Rabe, M., Figueira, G.: A review of simheuristics: extending metaheuristics to deal with stochastic optimization problems. Oper. Res. Perspect. **2**, 62–72 (2015)
13. Kizys, R., Juan, A., Sawik, B., Linares, A.: Solving the portfolio optimization problem under realistic constraints. In: Proceedings of the ICRA6/Risk 2015 International Conference, pp. 457–464, Barcelona, Spain (2015)
14. Bianchi, L., Marco, D., Gambardella, L.M., Gutjahr, W.J.: A survey on metaheuristics for stochastic combinatorial optimization. Nat. Comput. **8**, 239–287 (2009)
15. Chang, T.-J., Meade, N., Beasley, J.E., Sharaiha, Y.M.: Heuristics for cardinality constrained portfolio optimization. Comput. Oper. Res. **27**, 1271–1302 (2000)

The LIBOR Market Model Calibration
with Separated Approach

María Teresa Casparri[(⊠)], Martín Ezequiel Masci,
and Javier García-Fronti

Centro de Investigación en Métodos Cuantitativos Aplicados
a la Economía y la Gestión, IADCOM, Facultad de Ciencias Económicas,
Universidad de Buenos Aires, Av. Córdoba 2122, C1120AAQ CABA, Argentina
casparri@econ.uba.ar,
{martinmasci,javier.garciafronti}@economicas.uba.ar

Abstract. From an economic perspective, interest rates constitute key tools for decision making in the financial sector as they have micro and macro impacts, making its risk management a crucial matter. The LIBOR Market Model (LMM) uses the yield curve of the British interbank rate LIBOR (forward) as its basic input. Unlike models that use instantaneous rates, those involved in the LMM are observable in the market. Furthermore, the model is consistent with and adjusts its parameters according to the option valuation on futures formula in the Black '76 fashion. This allows for efficient calibration and can be used to value various derivative financial instruments. While there are several approaches for calibration, this work uses the *separated* approach with *optimization*. It is implemented using a routine in MATLAB with data of european swaptions. This work concludes that the proposed algorithm is computationally efficient and the fit is satisfactory.

Keywords: Interest rate model · Optimization algorithm · Financial derivatives valuation · Swaptions

1 Introduction

In order to illustrate the application of theoretical concepts developed in the LMM model[1], it is interesting to use swaptions on LIBOR [1]. This paper uses LIBOR market data to calibrate the LMM model by the optimized separation approach [2] and implement it in MATLAB. Model calibration consists in playing the market value of plain vanilla options [3].

Implementation poses a challenge which consists in transforming the unobservable values (such as forward rates and the rates volatility) into dynamics on observables [4]. Firstly, the parameters are set by minimizing the quadratic residues of the difference between the theoretical model and the pre-valuation of the derivatives in question. This procedure requires from the analyst to have discretion regarding the data to be used and, therefore, market information needs to be available [5]. The remaining parts of the model's implementation process, are post-calibration which it reflects the use of the

---

[1] For a more detailed development about stochastic calculus, see [1].

© Springer International Publishing Switzerland 2016
R. León et al. (Eds.): MS 2016, LNBIP 254, pp. 12–21, 2016.
DOI: 10.1007/978-3-319-40506-3_2

results, either for valuation of derivatives [6], or the associated cash flow, and in a context of less uncertainty.

The objective of this work[2] is to calibrate the LMM model using the separated approach and it is divided into two sections. The first one explains the LMM model and how to calibrate it. The latter applies this methodology to a database and calculates the parameters of the model in this context. After this, a conclusion is given, and future research is proposed.

2 The Model

Calibration of the LMM model implies finding parameters σ_i, $i = 1, \ldots, N$. This parameters represent the volatility of certain derivative. In this research paper, swaptions on LIBOR rate agreements are used [6, 7]. For MATLAB, this procedure can take between less than a minute and 15 min, depending on both the functional forms, and the number of iterations of the process. It is important to note that the process of calibrating is not something that should be done only once. An investor who uses these techniques to enhance their portfolio, should additionally think about how often it needs to re-calibrate the model [8, 9].

This section will be exposing some theoretical guidelines that justify the use of this approach for the calibration of the LMM model. This guidelines are taken from the book of Gatarek et al. [10] in which the model is studied in great detail. First, it is necessary to define the approach and the steps to outline the algorithm, which is then translated into computer language using MATLAB (2007a version).

As indicated by the authors mentioned, the algorithm belongs to the class of non-parametric calibration. The use of a matrix of volatilities allows for a straightforward implementation. The aim is to construct a covariance matrix forward LIBOR rate. To do this it will be necessary to compute different variants of the same approach depending on the set of parameters to be estimated: Λ_i. The relevant parameters are an array of lambdas associated with swaptions with various maturities [11]. An intermediate step is the construction of the covariance matrix (VCV) forward LIBOR rate by use of eigenvalues and eigenvectors. Here, sub-routines are necessary to correct data errors, so that we can assure the VCV is a positive definite matrix form. After this, one may proceed to minimize the mean square error between the theoretical valuation and market values.

The process creating a matrix of volatilities of swaptions [10] begins:

$$\sum_{SWPT} = \begin{bmatrix} \sigma_{1,2}^{swpt} & \sigma_{1,3}^{swpt} & \sigma_{1,4}^{swpt} & \cdots & \sigma_{1,m+1}^{swpt} \\ \sigma_{2,2}^{swpt} & \sigma_{2,4}^{swpt} & \sigma_{2,5}^{swpt} & \cdots & \sigma_{2,m+2}^{swpt} \\ \sigma_{3,4}^{swpt} & \sigma_{3,5}^{swpt} & \sigma_{3,6}^{swpt} & \cdots & \sigma_{3,m+3}^{swpt} \\ \cdots & \cdots & \cdots & \cdots & \cdots \\ \sigma_{m,m+1}^{swpt} & \sigma_{m,m+2}^{swpt} & \sigma_{m,m+3}^{swpt} & \cdots & \sigma_{m,M}^{swpt} \end{bmatrix}_{m \times m} \qquad (1)$$

[2] We appreciate the comments and help from Pablo Matías Herrera and Joaquín Bosano.

Each matrix component, $\sigma_{i,j}^{swpt} = \sigma^{swpt}(t, T_i, T_j)$ is the volatility of a *swaption* with *maturity* T_i (assuming an underling swap T_i, T_j).

After that, it is possible to define the covariance matrix for the forward LIBOR as:

$$\Phi^i = \begin{bmatrix} \varphi_{1,1}^i & \varphi_{1,2}^i & \varphi_{1,3}^i & \cdots & \varphi_{1,m}^i \\ \varphi_{2,1}^i & \varphi_{2,2}^i & \varphi_{2,3}^i & \cdots & \varphi_{2,m}^i \\ \varphi_{3,1}^i & \varphi_{3,2}^i & \varphi_{3,3}^i & \cdots & \varphi_{3,m}^i \\ \cdots & \cdots & \cdots & \cdots & \cdots \\ \varphi_{m,1}^i & \varphi_{m,2}^i & \varphi_{m,3}^i & \cdots & \varphi_{m,m}^i \end{bmatrix}_{m \times m} \tag{2}$$

<u>Source</u>: Gatarek et al. [10].

Where

$$\varphi_{kl}^i = \int_0^{T_i} \sigma^{inst}(t, T_{l-1}, T_l) \sigma^{inst}(t, T_{k-1}, T_k) dt; \quad i < k \wedge i < l \tag{3}$$

And $\sigma^{inst}(t, T_{l-1}, T_l)$ is the instant volatility of the LIBOR rate $L_l(t, T_{l-1}, T_l)$.

It is assumed that parameters Λ_i exist, such that:

$$\varphi_{kl}^i = \Lambda_i \varphi_{kl} \tag{4}$$

Assuming that $\Lambda_i = \delta_{0,k} \quad \forall k = 1, \ldots, m$, it is possible to calculate parameters of the principal diagonal of $\Phi_{m \times m}$:

$$\varphi_{kk} = \frac{\delta_{0,k} \cdot \sigma^{swpt}(t, T_k, T_{k+1})^2}{\Lambda_k} \tag{5}$$

The next step is to define parameters $R_{i,j}^k(t)$

$$R_{i,j}^k(t) = \frac{B(0, T_{k-1}) - B(0, T_k)}{B(0, T_i) - B(0, T_j)} \tag{6}$$

Where $B(0, T_n) \ n = 1, \ldots, M$ are the LIBOR discount factors, and vector B is defined as:

$$B = \begin{pmatrix} B(0, T_1) \\ \cdots \\ B(0, T_M) \end{pmatrix} \tag{7}$$

Please note that $R_{i,j}^k(t)$ depends on k, i, j and on the maturities selected in the calibration.

With these new definitions it is now possible to calculate the rest of matrix:

$$\varphi_{k,N-1} =$$
$$= \frac{\delta_k \cdot \sigma_{k,N}^2 - \Lambda_k \left(\sum_{l=k+1}^N \sum_{i=k+1}^N R_{k,N}^i(0) \cdot \varphi_{i-1,l-1} \cdot R_{k,N}^l(0) - 2 \cdot R_{k,N}^{k+1}(0) \cdot \varphi_{k,N-1} \cdot R_{k,N}^N(0) \right)}{2 \cdot \Lambda_k \cdot R_{k,N}^{k+1}(0) \cdot R_{k,N}^N(0)}$$

Gatarek et al. [10] use the model proposed by Longstaff et al. [12] to give a numerical solution. Initially the authors proposed the creation of a new covariance matrix called Φ^M, similar to the original matrix but without the eigenvectors associated with negative eigenvalues.

This procedure consists in multiplying each eigenvectors (e) by the square root of the associated eigenvalue ($\sqrt{\lambda_i}$). This is only done for positive λ_i. Then, a matrix is built with this changes. The matrix Φ^M is then, the result of the product between the modificated Φ^i by its transpose. This procedure fixes any problem related to negative values.

The development of this sub algorithm allows us to build the swaptions' theoretical values matrix by approaching their volatilities using, initial values of Φ^M.

Each component follows Gatarek et al. [10]:

$$\varphi_{kl}^{M^i} = \Lambda_i \varphi_{kl}^M$$

$$\delta_k \sigma_{k,N}^2 \cong \Lambda_k \sum_{l=k+1}^N \sum_{i=k+1}^N R_{k,N}^i(0) \cdot \varphi_{i-1,l-1}^M \cdot R_{k,N}^l(0) \tag{8}$$

Since $\delta_k = \Lambda_k$ (Logstaff-Schwartz-Santa Clara model), then:

$$\sigma_{k,N}^2 \cong \sum_{l=k+1}^N \sum_{i=k+1}^N R_{k,N}^i(0) \cdot \varphi_{i-1,l-1}^M \cdot R_{k,N}^l(0) \tag{9}$$

Towards this assumptions, it is possible to build market volatilities and address the accuracy by the root mean square error between theoretical assumptions and market values:

$$RMSE = \sum_{i,j=1}^m \left(\sigma_{ij}^{Theoretical} - \sigma_{ij}^{Market} \right)^2 \tag{10}$$

With this expression described above, the separated approach can be optimized by nonlinear functions. RSME admits minimum only if VCV is defined positive. The theoretical model explains better the reality of the market if minimization is made. In the following section, the developed algorithm is shown with real data using MATLAB.

3   Calibration for European Swaptions

First of all, in order to implement LMM with real data it should be taken into consideration three steps: calibration, derivative payoff and cash flows related to the option. This work examines the procedure to obtain the parameters needed in the first step.

In the MATLAB routine it is possible to program a recursive process in six steps. The recursive character is configured by the fact that each step needs information of previous steps. The next figure shows all the process, data input and output required (Fig. 1):

Step	Inputs	Outputs
1	Vector of Discount Factors **[B]**	The Matrix of Parameters **[R]**
2	i. **[R]** ii. Vector of Date **[T_num]** iii. Matrix of market swaption volatilities **[Sig]** iv. Vector of initial parameters **[Lambda]**	The matrix of covariances **[VCV]** as a function of parameters **[Lambda]**
3	**[VCV]**	i. The vector of EigenValues **[L]** as a function of parameters [Lambda] ii. The parix of eigenvestors **[E]** as a function of parameters [Lambda]
4	i. **[L]** ii. **[E]**	The modified covariance matrix **[VCV_M]**
5	i. **[R]** ii. **[VCV_M]**	Theoretical swaption volatilities **[Sig_theo]**
6	i. **[Sig_theo]** ii. **[Sig]**	RSME between theoretical and market swaption volatilities

Fig. 1. Algorithm – separated approach. (Source: compiled by authors based on Gatarek et al. [10])

As it was mentioned above, the recursive character is illustrated by the dotted arrow. The output data in each step is treated as input data of the next one.

This process is known as the separated approach and some initial data is necessary in order to execute the algorithm[3]: vector of dates, swaption market volatilities matrix and parameters vector (all belonging to step 2). Regarding to the first one, it is a column vector with schedule data according to the instruments tenors. If those *maturities* are expressed in years, vector has M dates of the same initial date but within a year difference. This paper is based on Gatarek et al. [10] data (European Swaptions from January 2005).

---

[3] A discount factor vector needs to be specified as well, this is easily calculated from interest rates and tenors.

As initial data it is necessary to put on a parameter vector: Λ_i. Taking into consideration this performing issues:

i. If an arbitrary vector is imposed, the MATLAB code will return an RSME that matches with this initial condition. Hardly this RSME will be the local minimum for all the problem.

ii. Rather than including an arbitrary vector *[Lambda]*, an optimization in which each component of the vector are control variables of the mentioned problem should be performed.

This is why the separated approach with optimization is clearly superior than specifying an arbitrary vector. An initial condition is used as a vector "Lambda0" which components goes between 1 to 10 (in this case). In MATLAB it is allowed a function fminsearch to solve the mean square error as a nonlinear optimization:

```
options=optimset('MaxIter',1000);
Lambda0=[1 2 3 4 5 6 7 8 9 10];
[Lambda,f]=fminsearch(@FunciónObjetivo,Lambda0,opti
ons);
```

Then, routine will run an optimization performance with some objective values. The output it yields is the minimized RSME and an initial parameters vector which matches with this minimized error. Syntax, in this case, has RSME as objective function in its arguments. Initial parameters vector (Λ_i) is a constraint.

At the same time, since the process consists of a minimization with a searching objective, it is possible to set a specific number of iterations to generate a faster solution. This paper concludes that 10.000 iterations are enough to find a local minimum in our problem. To prove this affirmation, the program was run without any iterations imposed. Results were the same in both, the restricted and the unrestricted cases.

Since Fig. 2, step 3 generates the eigenvalues vector and eigenvectors matrix as Λ_i functions (Fig. 3):

Eigenvectors (L):

$$
L =
\begin{bmatrix}
-0.0147 \\
0.0006 \\
0.0179 \\
0.0251 \\
0.0339 \\
0.0506 \\
0.0727 \\
0.1107 \\
0.1547 \\
0.2447
\end{bmatrix}
$$

VCV=

k/l (years)	1	2	3	4	5	6	7	8	9	10
1	0,0312	0,0109	0,0044	0,0022	0,0012	0,0007	0,0005	0,0003	0,0002	0,0002
2	0,0109	0,0454	0,0256	0,0102	0,0051	0,0029	0,0018	0,0012	0,0008	0,0006
3	0,0044	0,0256	0,0528	0,0377	0,0149	0,0075	0,0042	0,0027	0,0018	0,0012
4	0,0022	0,0102	0,0377	0,0545	0,0489	0,0194	0,0097	0,0056	0,0035	0,0024
5	0,0012	0,0051	0,0149	0,0489	0,0679	0,0462	0,0183	0,0095	0,0053	0,0035
6	0,0007	0,0029	0,0075	0,0194	0,0462	0,0656	0,0587	0,0236	0,0119	0,0068
7	0,0005	0,0018	0,0042	0,0097	0,0183	0,0587	0,0638	0,0695	0,0274	0,014
8	0,0003	0,0012	0,0027	0,0056	0,0095	0,0236	0,0695	0,0835	0,0582	0,0234
9	0,0002	0,0008	0,0018	0,0035	0,0053	0,0119	0,0274	0,0582	0,0959	0,0535
10	0,0002	0,0006	0,0012	0,0024	0,0035	0,0068	0,014	0,0234	0,0535	0,1359

Fig. 2. The matrix of covariances [VCV] as a function of parameters (Logstaff-Schwartz-Santa Clara model)

E=

i	e_1i	e_2i	e_3i	e_4i	e_5i	e_6i	e_7i	e_8i	e_9i	e_10i
1	-0,0006	0,0213	-0,267	0,6756	-0,6152	-0,2504	-0,1615	-0,0583	0,0336	0,0107
2	0,0068	-0,1474	0,4869	-0,3956	-0,3148	-0,4639	-0,4567	-0,208	0,1296	0,0414
3	-0,0385	0,4361	-0,469	-0,0618	0,3768	-0,1297	-0,4712	-0,3531	0,2655	0,0923
4	0,1045	-0,6733	0,0558	0,2668	0,2121	0,2931	-0,1007	-0,3716	0,394	0,1646
5	-0,223	0,4878	0,3641	0,0431	-0,2504	0,3201	0,3005	-0,2474	0,4546	0,2337
6	0,4932	-0,0833	-0,371	-0,3318	-0,1948	-0,2442	0,3881	0,116	0,3601	0,3329
7	-0,6891	-0,1889	-0,064	0,0496	0,1717	-0,323	0,0691	0,3643	0,1652	0,4292
8	0,4525	0,2151	0,3562	0,3145	0,2348	0,0112	-0,2793	0,3958	-0,0495	0,4805
9	-0,1207	-0,0849	-0,2517	-0,3015	-0,3655	0,5259	-0,3248	0,0289	-0,3142	0,4569
10	0,018	0,016	0,0638	0,0882	0,1292	-0,2761	0,3226	-0,5707	-0,54	0,4162

Fig. 3. Eigenvectors

Regarding the filter described above, it is possible to take away all eigenvectors associated with negative eigenvalues. In this case, only the first eigenvalue is negative. The VCV_M matrix or Φ^M is shown in Fig. 4:

VCV_M=

(years)	1	2	3	4	5	6	7	8	9	10
1	0,0312	0,0109	0,0044	0,0022	0,0012	0,0007	0,0005	0,0003	0,0002	0,0002
2	0,0109	0,0454	0,0256	0,0102	0,0051	0,0029	0,0017	0,0012	0,0008	0,0006
3	0,0044	0,0256	0,0528	0,0376	0,015	0,0072	0,0046	0,0024	0,0018	0,0012
4	0,0022	0,0102	0,0376	0,0547	0,0485	0,0202	0,0087	0,0063	0,0033	0,0024
5	0,0012	0,0051	0,015	0,0485	0,0686	0,0446	0,0206	0,008	0,0057	0,0034
6	0,0007	0,0029	0,0072	0,0202	0,0446	0,0692	0,0537	0,0269	0,011	0,007
7	0,0005	0,0017	0,0046	0,0087	0,0206	0,0537	0,0708	0,0649	0,0287	0,0138
8	0,0003	0,0012	0,0024	0,0063	0,008	0,0269	0,0649	0,0866	0,0574	0,0235
9	0,0002	0,0008	0,0018	0,0033	0,0057	0,011	0,0287	0,0574	0,0961	0,0535
10	0,0002	0,0006	0,0012	0,0024	0,0034	0,007	0,0138	0,0235	0,0535	0,1359

Fig. 4. The modified covariance matrix [VCV_M] as a function of parameters

It is possible to generate theoretical volatilities matrix according to the last matrix and [R] matrixes calculated in the beginning of the algorithm. This is shown in formula (9).

Taking into account all the important information, the following outputs show theoretical swaption volatilities, the observables one (market information) and their differences (Figs. 5 and 6):

Sig_theo=

Theoretical	1	2	3	4	5	6	7	8	9	10
1	22,70%	20,29%	19,96%	19,81%	19,94%	19,57%	19,24%	19,17%	18,89%	18,63%
2	22,40%	20,34%	19,26%	18,83%	18,06%	17,47%	17,19%	16,78%	16,42%	
3	20,90%	19,45%	18,70%	17,55%	16,72%	16,28%	15,76%	15,32%		
4	19,53%	19,61%	18,03%	16,93%	16,36%	15,69%	15,14%			
5	18,30%	16,64%	15,57%	14,96%	14,23%	13,64%				
6	17,93%	16,95%	16,34%	15,33%	14,54%					
7	17,61%	17,74%	16,45%	15,44%						
8	16,27%	15,07%	14,07%							
9	15,26%	14,28%								
10	14,50%									

Fig. 5. Calculation of theoretical swaption volatilities [Sig_theo]

Sig=

Market	1	2	3	4	5	6	7	8	9	10
1	22,70%	23,00%	22,10%	20,90%	19,60%	18,60%	17,60%	16,90%	16,30%	15,90%
2	22,40%	21,50%	20,50%	19,40%	18,30%	17,40%	16,70%	16,20%	15,80%	
3	20,90%	20,10%	19,00%	18,00%	17,00%	16,30%	15,80%	15,50%		
4	19,50%	18,70%	17,70%	16,80%	16,00%	15,50%	15,10%			
5	18,20%	17,40%	16,50%	15,80%	15,10%	14,80%				
6	17,46%	16,74%	15,90%	15,24%	14,62%					
7	16,72%	16,08%	15,30%	14,68%						
8	15,98%	15,42%	14,70%							
9	15,24%	14,76%								
10	14,50%									

Fig. 6. Matrix of market swaption volatilities [Sig]. (Source: compiled by authors based on Gatarek et al. [10]).

Hence, RSME between theoretical and market swaption volatilities (Eq. (10)) is:

$$\mathbf{RSME} = \sum_{i,j=1}^{m} \left(\sigma_{ij}^{Theoretical} - \sigma_{ij}^{Market}\right)^2 = \mathbf{0.005336}$$

After running the program, initial parameters vector (control variables of the problem) are obtained:

Lambda =
$$(\mathbf{1.6495 \quad 2.2113 \quad 2.4858 \quad 2.7925 \quad 2.4413 \quad 2.7904 \quad 3.0697 \quad 2.4470}$$
$$\mathbf{2.1824 \quad 1.5488})$$

As was described, this outcome results in an optimization that matches RSME with the corresponded parameters. This information is an input to valuate the financial derivative and its future prices curve. If the proposed algorithm is computationally efficient and the fit is satisfactory (in the RSME sense), this mentioned future prices curve will be certainly close to the actual future prices. Algorithm complexity has sensitivity to data input and, specifically, to the derivatives considered. The higher the tenor considered, and the lesser partial autocorrelation has the time series, the lesser will be the prediction power of the parameters found.

It is possible to run the program with 100 and 1.000 iterations. The minimum reached is better when iterations increase. In the first case, RSME is 0.091659, while in second case 0.008107.

$$0.091659 > 0.008107 > 0.005336$$
$$\underbrace{\qquad}_{RSME_{100}} \qquad \underbrace{\qquad}_{RSME_{1.000}} \qquad \underbrace{\qquad}_{RSME_{10.000}}$$

A RSME closer to zero means that LMM fits better to market data. A perfect fit is hard to achieve and has no sense to several financial analysts since reality has uncertainty and risks [13]. There is a challenge in reducing uncertainty of future events and using financial models for portfolio rebalance or creating value through investment strategies.

4 Conclusion

The LMM model aims to minimize root mean squared errors between the theoretical value and market data of European swaptions. This work analysed a calibration of this model using the methodology known as *separated approach with optimization* and implemented it in MATLAB.

This is especially important to understand the dynamics of the financial markets. In this complex environment, future is unpredictable, but investors could limit their risk exposure implementing routines like the one shown in this paper. The objective from the point of view of financial engineering is to assess the discrepancy between the market value and the theory of a financial asset, and minimize it using the "root mean squared error" methodology. Moreover, an investor who uses these techniques must, additionally, reflect on how often it should re-calibrate the model.

This work concluded that the proposed algorithm is computationally efficient and the fitting is appropriate. Particularly, the results of the non-linear minimization are sensitive to the number of iterations specified. Real-time valuation compromises its accuracy. Moreover, this procedure is not appropriate for valuing complex portfolios that include exotic derivatives due to the computational power needed for this type of valuations.

Future research aims to develop a framework for investors to evaluate different financial derivatives using LMM and also expand research on how to accelerate the algorithm presented.

References

1. Masci, M.E.: Cálculo Estocástico y Calibración en Modelos de Tasa de Interés: Aplicación al Mercado LIBOR. Masters degree thesis, Facultad de Ciencias Económicas, Universidad de Buenos Aires (2015)
2. Brace, A., Gatarek, D., Musiela, M.: The market model of interest rate dynamics. Math. Financ. **7**, 127–155 (1997)
3. Löeffler, G., Posch, P.N.: Credit Risk Modeling Using Excel and VBA. Wiley, London (2007)
4. Glasserman, P.: Monte Carlo Methods in Financial Engineering, vol. 53. Springer Science & Business Media, New York (2003)

5. Korn, R., Liang, Q.: Robust and accurate Monte Carlo simulation of (cross-) Gammas For Bermudan swaptions in the LIBOR market model. CFin. **17**, 87 (2014)
6. Andersen, L.B.G.: A simple approach to the pricing of Bermudan swaptions in the multi-factor LIBOR market model. SSRN 155208 (1999)
7. Bessis, J.: Risk Management in Banking. Wiley, West Sussex (2011)
8. Andersen, L., Piterbarg, V.: Interest Rate Modelling Volume 1: Foundations and Vanilla Models. Atlantic Financial Press, London (2010)
9. Andersen, L., Piterbarg, V.: Interest Rate Modelling Volume 2: Term Structure Models. Atlantic Financial Press, London (2010)
10. Gatarek, D., Bachert, P., Maksymiuk, R.: The LIBOR Market Model in Practice. Wiley, London (2007)
11. Leippold, M., Stromberg, J.: Time-changed Lévy LIBOR market model: pricing and joint estimation of the cap surface and swaption cube. JFE **111**, 224–250 (2014)
12. Longstaff, F.A., Santa-Clara, P., Schwartz, E.S.: Throwing away a billion dollars: the cost of suboptimal exercise strategies in the swaptions market. JFE **62**, 39–66 (2001)
13. Crouhy, M., Galai, D., Mark, R.: The Essentials of Risk Management. McGraw-Hill, New York (2006)

Solving Realistic Portfolio Optimization Problems via Metaheuristics: A Survey and an Example

Jana Doering[1(✉)], Angel A. Juan[2], Renatas Kizys[3], Angels Fito[1], and Laura Calvet[2]

[1] Economics and Business Department, Universitat Oberta de Catalunya, Barcelona, Spain
{jdoering, afitob}@uoc.edu
[2] Computer Science Department, Universitat Oberta de Catalunya – IN3, Castelldefels, Spain
{ajuanp, lcalvet1}@uoc.edu
[3] Subject Group Economics and Finance, University of Portsmouth, Portsmouth, UK
renatas.kizys@port.ac.uk

Abstract. Computational finance has become one of the emerging application fields of metaheuristic algorithms. In particular, these optimization methods are quickly becoming the solving approach alternative when dealing with realistic versions of financial problems, such as the popular portfolio optimization problem (POP). This paper reviews the scientific literature on the use of metaheuristics for solving rich versions of the POP and illustrates, with a numerical example, the capacity of these methods to provide high-quality solutions to complex POPs in short computing times, which might be a desirable property of solving methods that support real-time decision making.

Keywords: Metaheuristics · Financial applications · Portfolio optimization

1 Introduction

Since the last century the direct relationship between financial decisions and wealth creation through capital accumulation or economic development has been widely accepted and thus been the center of much research in modern academia. Most of the questions in financial economics are formulated as optimization problems. Traditionally, exact methods have been employed in determining optimal solutions. This, however, requires modeling the problem subject to strict assumptions and simplifying formulations to render it solvable. Because this approach neglects depicting the intricacies of the real-life problems that decision-makers face in their everyday actions, the results are largely not transferrable to operations without reservations. Furthermore, the extraordinary internationalization and integration of financial markets and institutions has caused the decision-making process to become more complex. Advances in operations research and computer sciences have brought forward new solution approaches in optimization theory, such as metaheuristics [28]. While simple heuristics

© Springer International Publishing Switzerland 2016
R. León et al. (Eds.): MS 2016, LNBIP 254, pp. 22–30, 2016.
DOI: 10.1007/978-3-319-40506-3_3

have been employed since the 1960s, the more advanced metaheuristics have only recently become relevant with the increase in computing power. They have shown to provide solutions to problems for which traditional methods are not applicable [22] in addition to providing near-optimal solutions to complex-combinatorial-optimization problems. Metaheuristics are conceptually simple, relatively easy to implement, and usually require little computational time, making them attractive for problem solving in applications requiring real-time decisions. The application of metaheuristics to financial problems is summarized in Gilli et al. [9]. While metaheuristics do not guarantee finding a global optimal solution, Gilli and Schumann [10] point out that the goal of optimization in most real-life instances is not to provide an optimal solution, but one that fulfills the objectives to a satisfactory extent and hence promote the use of heuristic approaches. With respect to exact methods that provide an optimal solution to approximated problem formulations, a near-optimal solution to the unrestricted problem combined with real-life constraints might be preferable.

Since Markowitz [20] developed the modern portfolio theory centered on the mean-variance approach the academic community has been highly engaged in advancing the tools for portfolio optimization. The theory is based on two constituting assumptions, namely the financial investor being concerned with the expected returns and the risk of their respective investment. It is thus the goal to minimize the level of risk expressed through the portfolio variance for a given expected return level, resulting in the unconstrained efficient frontier. This established the portfolio optimization problem (POP), in which the risk is sought to be minimized based on a minimum expected return required by the investor. Figure 1 shows the increasing popularity of metaheuristics for solving the POP. It becomes obvious that the trend in publications is increasing and that metaheuristics have received increased attention as solution approaches in the area of portfolio optimization.

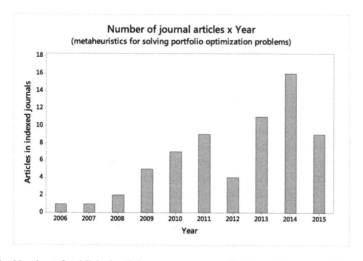

Fig. 1. Number of published articles per year on applications of metaheuristics in POP

While Mansini et al. [18] provide an extensive review on different portfolio optimization problems, including its historical evolution and the use of exact methods, our paper focuses on more recent contributions of metaheuristics for solving realistic versions of the POP, both including single-objective and multi-objective optimization. This paper also includes a computational example that illustrates the potential of metaheuristics in the field.

2 The Single-Objective Portfolio Optimization Problem

While the original Markowitz problem can be solved using quadratic programming, metaheuristics have increasingly been employed to cope with the fact that the problem becomes NP-hard when real-life constraints are introduced [3]. These constraints include cardinality constraints (restricting the number of assets in the portfolio) and minimum proportional restrictions for inclusion of any asset. The classical version of the POP can be considered a single objective optimization problem with either one of the following model formulations: The investor minimizes the risk exposure subject to a minimum attainable expected return or the investor maximizes the expected return for a given level of risk. The classical POP can be formulated as follows [4]. The risk expressed as the portfolio variance is minimized:

$$\sum_{i=1}^{N} \sum_{j=1}^{N} w_i * w_j * \sigma_{ij}, \tag{1}$$

subject to a minimum return, the constraint that the weights have to add up to one and the constraint that all asset weights must lie between zero and one, inclusive.

$$\sum_{i=1}^{N} w_i * \mu_i \geq R^*, \tag{2}$$

$$\sum_{i=1}^{N} w_i = 1, \tag{3}$$

$$0 \leq w_i \leq 1; \quad i = 1\ldots1, N, \tag{4}$$

where N is the total number of assets available, μ_i is the expected return of an asset i, R^* is the minimum required return, w are the respective weights of the assets making up the portfolio and σ_{ij} is the covariance between two assets i and j.

Chang et al. [4] solve the above classical problem definition using three different metaheuristic approaches in order to generate a cardinality-constrained efficient frontier: genetic algorithm (GA), tabu search (TS), and simulated annealing (SA). They suggest pooling the results from the different approaches because no single heuristic was uniformly dominating in all observed datasets. Following this suggestion and combining GA, TS, and SA, Woodside-Oriakhi et al. [31] further explore the pooling option. They find that, on average, SA contributes little to the performance of the process and that thus a pooled GA and TS heuristic is superior to single metaheuristic approaches at the expense of higher computational time. By combining exact and metaheuristic methods, they create matheuristics. However, Soleimani et al. [26]

introduce sector capitalization and minimum transaction lots as further constraints and found that the GA they developed outperformed TS, and SA. Particle swarm optimization (PSO) was found to be competitive with all three, GA, TS, and SA for the cardinality-constrained portfolio selection problem and especially successful in low-risk portfolios [5]. Golmakani and Fazel [11] further introduce minimum transaction lots, bounds on holdings and sector capitalization in addition to cardinality constraints and apply a combination of binary PSO and improved PSO that they call CBIPSO and found that, especially for large-scale problems, CBIPSO outperforms GA in that it provides better solutions in less computing time.

Di Tollo and Roli [8] provide a survey concerned with the early applications of metaheuristics to the POP and some of the proposed constraints. They explicitly highlight the potential use of hybrid approaches. Such a hybrid method was proposed by Maringer and Kellerer [19] who employ hybrid local search algorithm, which combines principles of SA and evolutionary algorithms (EA), to optimize a cardinality-constrained portfolio. The option of hybrid approaches is further investigated by Di Gaspero et al. [7] who combine a local search metaheuristics with quadratic programming to optimize a portfolio while accounting for cardinality constraints, lower and upper boundaries for the quantity of an included asset and pre-assignment constraints. According to their results, the developed solver finds the optimal solution in several instances and is at least comparable to other state of the art methods.

3   The Multi-objective Portfolio Optimization Problem

While single objective optimization methods consider either a minimal risk for a given return or a maximum risk for a given expected return or an objective function that weights the two goals and thus have to be run several times with the respective weights [23], multi-objective optimization methods find a set of Pareto solutions, while balancing two or more objective functions simultaneously. According to Streichert et al. [27] the problem can then be formulated as follows. For a multi-objective optimization it becomes necessary to minimize the portfolio risk expressed by the portfolio variance:

$$\sum_{i=1}^{N} \sum_{j=1}^{N} w_i * w_j * \sigma_{ij}, \tag{5}$$

while maximizing the return of the portfolio

$$\sum_{i=1}^{N} w_i * \mu_i, \tag{6}$$

subject to

$$\sum_{i=1}^{N} w_i = 1, \tag{7}$$

$$0 \leq w_i \leq 1; \quad i = 1 \ldots 1, N, \tag{8}$$

Zhu et al. [32] aim at a comparison of GA and particle swarm optimization (PSO) in solving the non-linear portfolio optimization problem with multi-objective functions. They argue that PSO overcomes the problem of increased convergence time in large instances expected for GA. They test their methodology on unconstrained, as well as constrained portfolios. While they do not include constraints other than a total weight equal to one, in addition to restricted portfolios, in which the short-selling of the portfolio's underlying assets is prohibited and thus all asset weights are positive, the authors also investigate unrestricted portfolios. However, they introduce the Sharpe ratio as a simultaneous measure of risk and return and thus turn the multi-objective optimization problem into a single objective optimization by optimizing an objective function that serves as a simultaneous measure of risk and return. The solution portfolios obtained with the PSO solver outperformed those constructed using GA for all test problems in terms of Sharpe ratio and the established efficient frontier was above that of GA portfolios in all but one instance. Enhanced PSO algorithms for solving the multi-objective POP have been proposed by Deng et al. [6] and He and Huang [12]. Cardinality and bounding constraints are incorporated by Deng et al. [6], who find that their algorithm mostly outperforms GA, SA, and TS algorithms as well as previous PSO approaches especially in the case of low-risk portfolios. It can be concluded that different findings unanimously favor PSO in situations when low-risk investment is demanded. Similarly, He and Huang [12] propose a modified PSO (MPSO) algorithm that outperforms regular PSO for their four different optimization sets. These sets consist of the traditional Markowitz mean-variance formulation and three alternative discontinuous objective functions that simultaneously account for minimizing risk while maximizing returns. More recently, they also developed a new PSO to further enhance discontinuous modeling of the POP and find that it generally outperforms PSO and also performs better than MPSO in larger search spaces [13]. Other population-based algorithms applied in optimizing portfolios include firefly algorithms (FA) [29] and artificial bee colony (ABC) algorithms. The authors developed these to address unconstrained portfolio optimization as well as portfolios with cardinality and bounding constraints. However, because the results were satisfactory at most even after modifications, the authors hybridized FA and ABC by incorporating the FA's search strategy into ABC to enhance exploitation and found that their data suggested superiority of the methodology compared to GA, SA, TS, and PSO [30] for unconstrained and cardinality-constrained portfolios.

Streichert et al. [27] account for further constraints, namely buy-in thresholds (acquisition prices) and roundlots (smallest volume of an asset that can be purchased). They treat the POP as a multi-objective optimization problem, in which they simultaneously minimize risk while maximizing expected returns through two multi-objective evolutionary algorithms (MOEA): GA and evolutionary algorithm (EA) enhanced through the integration of a local search that applies Lamarckism. They found that this enhancement greatly improved the reliability of the results, especially with respect to the additional constraints. However, there is a second point of criticism to the original Markowitz model, namely its assumption of normal financial returns, which, in reality are characterized by a leptokurtic and fat-tailed distribution [15], making it necessary to consider non-parametric risk measures. Such a measure is the value-at-risk as employed by

Babaei et al. [2], who developed two multi-objective algorithms based on PSO to solve a cardinality- and quantity-constrained POP. Through splitting the whole swarm into sub-swarms that are then evolved distinctly their methodology outperformed similar benchmark metaheuristics. In order to optimize a non-parametric value-at-risk and to include further constraints, including a lower and upper bounds for the weights of included assets, a threshold for asset weight changes, lower and upper bounds for the weights of one asset class and a turnover rate that determines the maximum asset allocation changes possible at once, Krink and Paterlini [16] developed the differential evolution for multi-objective portfolio optimization (DEMPO) algorithm, partly based on differential evolution (DE). An extended version of a generalized DE metaheuristic is also employed in optimizing a highly constrained POP by Ayodele and Charles [1]. The included constraints consist of bounds on holdings, cardinality, minimum transaction lots, and expert opinion. An expert can form an opinion based on indicators beyond the scope of the analyzed data and influence whether or not an asset should be included. Their methodology shows improved performance when compared to GA, TS, SA, and PSO. Lwin et al. [17] considered cardinality, quantity, pre-assignment and round lot constraints and developed a multi-objective evolutionary algorithm that is improved through a learning-guided solution generation strategy, which promotes efficient convergence (learning-guided multi-objective evolutionary algorithm with external archive, MODEwAwL). It was shown that the developed algorithm outperforms four benchmark state of the art multi-objective evolutionary algorithms in that its efficient frontier is superior. An extensive review of the application of evolutionary algorithms is provided by Metaxiotis and Liagkouras [21].

4 A Numerical Example

In Kizys et al. [14], the authors describe a metaheuristic algorithm that allows to solve a realistic version of the POP. In order to test the efficiency of their approach, we have used their algorithm with a set of data retrieved from the repository ORlib (http://people. brunel.ac.uk/ ~ mastjjb/jeb/orlib/portinfo.html). These instances were proposed by Chang et al. [4] and were studied by Schaerf [25], Moral-Escudero et al. [24], and Di Gaspero et al. [7]. The data set comprises constituents of five stock market indices, Hang Seng (Hong Kong), DAX 100 (Germany), FTSE 100 (United Kingdom), S&P 100 (United States) and NIKKEI 225 (Japan). Following Di Gaspero et al. [7] we divided the portfolio frontier into 100 equidistant points on the vertical axis that represents the user-defined rate of portfolio expected return. The algorithm was executed on a constrained mean-variance frontier. The benchmark constraints are those imposed by the previous authors. Figure 2 shows the values of average percentage loss (APL) and associated computational times for several algorithms. Notice that, for most instances, the algorithm from Kizys et al. [14] is able to outperform, in terms of the minimum APL (the lower the better), the algorithms proposed by Moral-Escudero et al. [24] and Di Gaspero et al. [7]. Also, the algorithm of the former authors shows its efficiency in terms of computational times requested to obtain a high-quality solution (Fig. 3).

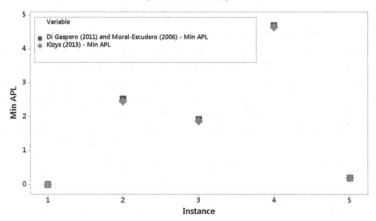

Fig. 2. Min APL (the lower the better) by instance and algorithm (Color figure online)

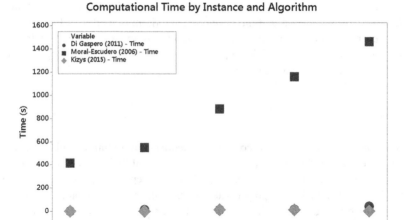

Fig. 3. Computational time by instance and algorithm (Color figure online)

5 Conclusions

This paper analyzes the role of metaheuristic-based approaches in solving realistic variants of the well-known portfolio optimization problem, either with single or with multiple objectives. As discussed in the paper, metaheuristic algorithms are gaining popularity to solve these rich variants of the problem, since they might be used even in those scenarios in which exact methods cannot provide optimal solutions in reasonable computing times. An example of application illustrates the efficiency of these algorithms, in particular their capacity to provide high-quality solution in short computing times.

Acknowledgments. This work has been partially supported with doctoral grants from the UOC, the Spanish Ministry of Economy and Competitiveness (grants TRA2013-48180-C3-3-P, TRA2015-71883-REDT) and FEDER. Likewise we want to acknowledge the support received by the Department of Universities, Research & Information Society of the Catalan Government (Grant 2014-CTP-00001).

References

1. Ayodele, A.A., Charles, K.A.: Portfolio selection problem using generalized differential evolution 3. Appl. Math. Sci. **9**(42), 2069–2082 (2015)
2. Babaei, S., Sepehri, M.M., Babaei, E.: Multi-objective portfolio optimization considering the dependence structure of asset returns. Eur. J. Oper. Res. **244**, 525–539 (2015)
3. Beasley, J.E.: Portfolio optimisation: models and solution approaches. In: Topaloglu, H., Smith, J.C. (eds.) 2013 Tutorials in Operations Research: Theory Driven by Influential Applications, pp. 201–221. INFORMS (2013)
4. Chang, T.J., Meade, N., Beasley, J.E., Sharaiha, Y.M.: Heuristics for cardinality constrained portfolio optimisation. Comput. Oper. Res. **27**, 1271–1302 (2000)
5. Cura, T.: Particle swarm optimization approach to portfolio optimization. Nonlinear Anal. Real World Appl. **10**, 2396–2406 (2009)
6. Deng, G.-F., Lin, W.-T., Lo, C.-C.: Markowitz-based portfolio selection with cardinality constraints using improved particle swarm optimization. Expert Syst. Appl. **39**, 4558–4566 (2012)
7. Di Gaspero, L., Di Tollo, G., Roli, A., Schaerf, A.: Hybrid metaheuristics for constrained portfolio selection problems. Quant. Financ. **11**, 1473–1487 (2011)
8. Di Tollo, G., Roli, A.: Metaheuristics for the portfolio selection problem. Int. J. Oper. Res. **6**, 13–35 (2008)
9. Gilli, M., Maringer, D., Schumann, E.: Numerical Methods and Optimization in Finance. Academic Press, Oxford (2011)
10. Gilli, M., Schumann, E.: Heuristic optimisation in financial modelling. Ann. Oper. Res. **193**, 129–158 (2012)
11. Golmakani, H.R., Fazel, M.: Constrained portfolio selection using particle swarm optimization. Expert Syst. Appl. **38**, 8327–8335 (2011)
12. He, G., Huang, N.: A modified particle swarm optimization algorithm with applications. Appl. Math. Comput. **219**, 1053–1060 (2012)
13. He, G., Huang, N.: A new particle swarm optimization algorithm with an application. Appl. Math. Comput. **232**, 521–528 (2014)
14. Kizys, R., Juan, A., Sawik, B. Linares, A.: Solving the portfolio optimization problem under realistic constraints. In: Proceedings of 2015 International Conference on ICRA6/Risk, pp. 457–464 (2015)
15. Krink, T., Paterlini, S.: Differential evolution and combinatorial search for constrained index-tracking. Ann. Oper. Res. **172**, 153–176 (2009)
16. Krink, T., Paterlini, S.: Multiobjective optimization using differential evolution for real-world portfolio optimization. Comput. Manag. Sci. **8**, 157–179 (2011)
17. Lwin, K., Qu, R., Kendall, G.: A learning-guided multi-objective evolutionary algorithm for constrained portfolio optimization. Appl. Soft Comput. **24**, 757–772 (2014)
18. Mansini, R., Ogryczak, W., Speranza, M.G.: Twenty years of linear programming based portfolio optimization. Eur. J. Oper. Res. **234**, 518–535 (2014)

19. Maringer, D., Kellerer, H.: Optimization of cardinality constrained portfolios with a hybrid local search algorithm. OR Spectr. **25**, 481–495 (2003)
20. Markowitz, H.M.: Portfolio selection. J. Financ. **7**, 77–91 (1952)
21. Metaxiotis, K., Liagkouras, K.: Multiobjective evolutionary algorithms for portfolio management: a comprehensive literature review. Expert Syst. Appl. **39**, 11685–11698 (2012)
22. Michalewicz, Z., Fogel, D.B.: How to Solve It: Modern Heuristics. Springer, Berlin (2004)
23. Mishra, S.K., Panda, G., Majhi, R.: Constrained portfolio asset selection using multiobjective bacteria foraging optimization. Oper. Res. **14**, 113–145 (2014)
24. Moral-Escudero, R., Ruiz-Torrubiano, R., Suárez, A.: Selection of optimal investment portfolios with cardinality constraints. In: IEEE Congress on Evolutionary Computation, pp. 2382–2388. IEEE Press (2006)
25. Schaerf, A.: Local search techniques for constrained portfolio selection problems. Comput. Econ. **20**, 177–190 (2002)
26. Soleimani, H., Golmakani, H.R., Salimi, M.H.: Markowitz-based portfolio selection with minimum transaction lots, cardinality constraints and regarding sector capitalization using genetic algorithm. Expert Syst. Appl. **36**, 5058–5063 (2009)
27. Streichert, F., Ulmer, H., Zell, A.: Evolutionary algorithms and the cardinality constrained portfolio optimization problem. In: Operations Research Proceedings 2003, Selected Papers of the International Conference on Operations Research (OR 2003), pp. 253–260 (2003)
28. Talbi, E.-G.: Metaheuristics: from Design to Implementation. Wiley, Hoboken (2009)
29. Tuba, M., Bacanin, N.: Artificial Bee Colony algorithm hybridized with Firefly algorithm for cardinality constrained mean-variance portfolio selection problem. Appl. Math. Inf. Sci. **8**, 2831–2844 (2014)
30. Tuba, M., Bacanin, N.: Upgraded Firefly algorithm for portfolio optimization problem. In: 2014 UKSim-AMSS 16th International Conference on Computer Modelling and Simulation, pp. 113–118 (2014)
31. Woodside-Oriakhi, M., Lucas, C., Beasley, J.E.: Heuristic algorithms for the cardinality constrained efficient frontier. Eur. J. Oper. Res. **213**, 538–550 (2011)
32. Zhu, H., Wang, Y., Wang, K., Chen, Y.: Particle swarm optimization (PSO) for the constrained portfolio optimization problem. Expert Syst. Appl. **38**, 10161–10169 (2011)

A Structural Equation Model for Analyzing the Association Between Some Attributes of Earnings Quality and Value Relevance. Results from the Structural Model

Cristina Ferrer[✉], Susana Callao, José Ignacio Jarne,
and José Antonio Laínez

University of Zaragoza, Zaragoza, Spain
{cferrer, scallao, jijarne, lainez}@unizar.es

Abstract. This study investigates the impact of earnings quality on the relationship between earnings and stock prices. In order to incorporate the multidimensional nature of earnings quality, we use Structural Equation Models. This methodology is suitable for mitigating the effects of choosing the measurement method of earnings quality and it has been highly developed in other social sciences fields, but it is underused in accounting research. Our results show that investors value the quality of accruals and dislike earnings manipulated by discretionary accruals, except when this manipulation reduces earnings volatility. The study is focused on various desirable features of earnings in order to evaluate the relationship between earnings quality and its value relevance. It values whether earnings management is opportunistic or informative from a new approach and, finally, it provides a major methodological contribution to the study of earnings quality with Structural Equation Models.

Keywords: Earnings quality · Value relevance · Discretionary accruals · Accruals quality · Earnings smoothing · Structural equation models

1 Introduction

The usefulness of financial reporting improves decision-making, but it is based on quality of accounting information. The analysis of this usefulness is usually focused on the quality of earnings because earnings are closely followed by market participants, provide useful information about the firm's valuation, are the best measure of a firm's performance and a good indicator of future cash flows (e.g. [1, 2]).

We concentrate on the usefulness of earnings for investment decisions studying the quality of earnings and the impact of that quality on stock pricing. Previous literature studying the capital market consequences of earnings quality focused on the cost of capital, trading volume or the information risk (e.g. [2, 3]). Other research studies the effect of earnings quality on stock prices (e.g. [4–7]).

In general, earnings quality is defined as a multidimensional concept and there is no an agreed-upon meaning assigned. In turn, empirical research has developed several metrics to proxy for earnings quality (see the surveys in, e.g. [1, 8]). The multidimensional

© Springer International Publishing Switzerland 2016
R. León et al. (Eds.): MS 2016, LNBIP 254, pp. 31–39, 2016.
DOI: 10.1007/978-3-319-40506-3_4

perspective of earnings quality makes it really difficult to measure [9] and some authors demonstrate that research of earnings quality cannot be limited to just one measure because these measures capture different aspects and may lead to contradictory conclusions [8–10].

In this context, the main motivation of our work is to analyze the relationship between earnings quality and capital market but introducing the multidimensional perspective of earnings quality. In order to answer to this motivation, this research is based on three desirable features of earnings as measures for earnings quality (discretionary accruals, accruals quality and earnings smoothing) and we apply a methodology that allows us to consider more than one alternative measurement: Structural Equation Models (SEM), which has been underused in accounting research.

Therefore, our objective is to analyze the impact of earnings quality on the relationship between earnings and stock prices. Investors follow earnings as an important variable for decision-making and, consequently, the quality of earnings should influence the relationship between earnings and stock prices.

Our work confirms that the results obtained from an earnings quality analysis are influenced by the method used to measure them. It also shows that the impact of earnings quality on the relation between earnings and stock prices is positive and significant. Discretionary accruals have a negative impact on this relationship. The more discretionary the accruals are the less relationship exists between earnings and stock prices. Accruals quality improves the relationship between earnings and stock prices. Earnings smoothing has a positive and significant impact; the more earnings are smoothed, the better they predict the market value of firms. These results provide evidence that investors value the quality of accruals and dislike earnings manipulated by discretionary accruals, except when this manipulation reduces the earnings volatility.

We make several contributions to the literature. Firstly, we focus on various desirable features of earnings and we evaluate how investors value each earnings quality measure. Apart from that, our study contributes to the literature by suggesting an alternative approach for valuing whether earnings management is opportunistic or informative. Finally, it provides a major methodological contribution to the study of earnings quality using structural equation models. This methodology is suitable for the study of aspects related to earnings quality allowing us to base the study on different models for measuring the same variable to mitigate the effects of choosing one single measurement method.

The remainder of the paper is organized as follows. The Sect. 2 reviews the previous literature and develops the hypotheses. The Sect. 3 describes the data and methodology. Then, we present the results obtained (Sect. 4) and the main conclusions (Sect. 5).

2   Background and Hypothesis Development

As we mentioned before, there is not an agreed-upon meaning assigned to "accounting quality". Empirical research defines it by very different types of measurements and, usually, considers just one dimension or one characteristic of earnings in isolation. Even so, at the beginning of the last decade, the multidimensional nature of earnings quality began to be incorporated into empirical research [1, 2].

Most studies on the capital market consequences of earnings quality focus on the cost of capital. Francis et al. [2] find a positive relationship between financial reporting quality and cost of equity in the US. Bhattacharya et al. [3] present country-level evidence; they find that an increase in earnings opacity is linked to an increase in the cost of equity and a decline in trading volume. Other studies have focused on the effect of earnings quality on stock prices. Subramanyam [4] finds that markets attach value to discretionary accruals and reports evidence that discretionary accruals predict future profitability and dividend changes. Chung et al. [5] find that discretionary accruals enhance the relationship between earnings and stock prices.

Based on previous literature, our objective is to find evidence regarding the effect of discretionary accruals, accruals quality, and earnings smoothing on the relationship between earnings and stock prices, in other words, between earnings quality and the value relevance of earnings, defined as the predicted association between earnings and equity market values. That is because the usefulness of financial information for making investment decisions is linked to relevance.

There seems to be a prevalent perception that earnings management is opportunistic in nature and, therefore, this manipulation reduces the value relevance of earnings. However, a number of academic studies have argued that earnings management may be beneficial whether it enhances the information value of earnings.

The opportunistic perception of earnings management increased after the scandals at Enron, Worldcom and elsewhere. The argument that earnings management decreases earnings quality is based on the asymmetric information. Managers intentionally modify earnings amounts and users do not have enough information to distinguish the data that has been altered [7, 11–13].

On the other hand, some researchers argue the informative nature of earnings management. It communicates private information about future expectations of performance, thereby facilitating the ability of investors to predict future results. Managers may exercise discretion over earnings to communicate private information to stockholders and the public [4, 5, 14–18].

Consequently, it is not easy to anticipate the effect of earnings management on value relevance and we have to consider both perspectives. *Discretionary Accruals* are an estimation of accruals that are not well explained by accounting fundamentals (fixed assets and revenues) being a measure of management's discretion over earnings.

Considering both perspectives and based on previous studies, we are unable to anticipate the sign of the relationship between discretionary accruals and the usefulness of earnings in determining stock prices. Therefore, we formulate our first hypothesis:

H1: There is a significant relationship between discretionary accruals and the value relevance of earnings as reflected in stock prices.

Accruals Quality captures the mapping of current accruals into last-period, current-period, and next-period cash flows. Accruals quality is based on the view that earnings that map closely to cash flows are of better quality [2, 10, 19–21].

Researchers have dedicated much effort to establishing links between earnings quality and security valuation, the cost of capital, and external financing decisions (see [8]). Taking into account the results of that previous research, which show that low earnings quality has a negative impact on these aspects, we formulate our second hypothesis as follow:

H2: There is a positive and significant relationship between accruals quality and the value relevance of earnings as reflected in stock prices.

Earnings Smoothing means a reduction in earnings variability, and implies the intentional nature of the smoothing. It indicates the attempt of managers to be discretionary in the publication of information so as to intentionally eliminate fluctuations from earnings [22–24].

Earnings smoothing has been the subject of extensive research and debate over the last few decades [4, 22, 25, 26] because it is seen from different perspectives in relation to the usefulness of financial reporting. Some authors think that earnings smoothing improves the informative content of earnings and, as such, is a desirable earnings feature [17, 25, 27–29]. Other authors consider that earnings smoothing introduces manipulation in earnings figures and indicates poor quality of accounting regulation because it allows managers to hide information from the users [30–32]. Nevertheless, the debate on the dominant effect of the earnings smoothing is still opened.

In view of the research results, we cannot anticipate if earnings are more relevant in determining stock prices in the presence of earnings smoothing or not. So, our third hypothesis is formulated as follows (in the alternative form):

H3: There is a significant relationship between earnings smoothing and the value relevance of earnings as reflected in stock prices.

3   Data and Methodology

The study is based on a sample of Spanish firms listed on the stock market. The period of analysis ran from 1998 to 2007. We employed the Amadeus and Compustat databases to obtain financial and market data, respectively. Firms belonging to financial and insurance sectors, those firms without information available for the minimum of 5 years and those observations referring to financial years shorter than one year were removed.

As we mentioned, we studied three earnings characteristics, derived from the relationships between accruals and cash: discretionary accruals, accruals quality, and earnings smoothing. All these earnings characteristics have also been assessed using different measures in order to build latent variables. For that, it is required to obtain earnings quality indicators for each firm. Therefore, based on Francis et al. [2], we calculated every earnings quality measure over firm-specific ten-year windows, from t-9 to t. It means that we must regress 11 different models to estimate discretionary accruals for each firm, to calculate 20 accruals quality indicators, and 15 different earnings smoothing ratios. All of them are based on previous literature because they have been measured in many different ways and our objective is to measure them by latent variables. All measures obtained for each of the attributes and their descriptions can be seen in [33, 34].

SEM is suitable for the study of earnings quality, measuring it through different variables that are not directly observable. These unobservable variables are considered as latent and this methodology allow us to measure earnings quality features based on the group of indicators that characterize them, by incorporating all the information that each of them provide.

PLS consists of two parts that are usually analyzed and interpreted sequentially in two stages: the measurement model and the structural model. The **measurement model** contains the relationships between each construct and its indicators. This step attempts to identify a set of measurable indicators that define the latent variable and is based on the study of the internal consistency of the constructs or latent variables. The **structural model** determines the causal relationships among the latent variables.

To consider a construct internally consistent, it has to fulfil the following properties: **unidimensionality**, that considers if the indicators that comprise each construct are one-dimensional, meaning that all of them measure just one dimension; **reliability**, that measures the consistency of the indicators that form the construct, whether or not the indicators are measuring the same concept; **convergent validity**, that is the degree to which the indicators reflect the construct, that is, whether they really measure what they purport to measure; and **discriminant validity**, that means that each construct must be significantly different from the remaining constructs to which it is not related. A construct satisfies this property when it has weak correlations with other latent variables.

By analyzing the fulfilment of these properties, we will be able to establish consistent measures of each earnings characteristic The evaluation of these properties makes up the exploratory analysis of the measurement model and, later, the final measurement model has also been confirmed in the confirmatory analysis through the PLS algorithm. Results for the measurement model can be seen in [33, 34].

On the other hand, to measure the relationship between earnings and stock prices, we evaluate the value relevance of earnings, considering earnings' ability to explain the firm's price on the stock market. Consistent with prior research, we measure value relevance by the R-squared of a model that establishes a relationship between market value and earnings [2], including both earnings changes and earnings levels. So, we estimate the following model for each firm over a period of ten years:

$$RET_{j,t} = \delta_{0,j} + \delta_{1,j}X_{j,t} + \delta_{2,j}\Delta X_{j,t} + \varepsilon_{j,t} \tag{1}$$

Where $RET_{j,t}$ is the firm j's 15-month market adjusted return ending three months after the end of fiscal year t; $X_{j,t}$ and $\Delta X_{j,t}$ are firm j's net income before extraordinary items (NIBE) in year t and its change between t and t−1, scaled by market value at the end of year t−1.

As we stated previously, the second stage is based on the analysis of the structural model which determines the causal relationships among the latent variables. The significance of all the paths of the structural model was tested using a bootstrap resampling procedure with 500 iterations in SmartPLS.

The path coefficients are the beta standardized regression coefficients (β) and they measure the relationship between two latent variables and its significance. The R-squares measure the proportion of the construct variance that is explained by the model.

4 Results

We have considered previously that value relevance of earnings may be the consequence of earnings quality measured through three different perspectives and we try to answer if it may influence the value relevance. We study the relationships proposed in Sect. 3 on the structural model and, in this section, we present the PLS results of testing the research propositions involved in the structural model and discusses the explanatory power of the model.

As we can see in Table 1, all the earnings quality measures, defined as the latent variables analyzed in the previous measurement model, show a significant relationship with value relevance.

Table 1. Results of structural model. Relationships between constructs

Hypothesis		Relationship	β	t value
H1	H1a	DA → VR	−0.2667	1.6693***
	H1b	AQ → VR	0.2784	2.5248**
	H1c	ESM → VR	0.4517	4.0367*

β: path coefficient, t-statistic of path coefficient
*significant at 1 %, **significant at 5 %, ***significant at 10 %

We begin with hypothesis H1, which tests if discretionary accruals (DA) can explain the value relevance of earnings on the stock market (VR). In this case, the results show a significant relationship of DA in the explanation of VR ($\beta = -0.2667$; $p < 0.10$). It can be seen that there is a negative relationship, which means that higher levels of discretionary accruals determine lower value relevance of earnings.

When we formulated this hypothesis, we could not anticipate the sign of the relationship because discretionary accruals can be evaluated from two perspectives: informative and opportunistic. According to these results, the capital market sees discretionary accruals as reflecting opportunistic conduct by managers who use their position within the company to influence the firm's financial reporting. This opportunistic earnings management impairs the value relevance of financial information. Discretionary accruals create distortions in the reported earnings and it reduces the ability of earnings to explain returns. The higher the level of discretionary accruals, the less informative the earnings announcement is for investors.

Hypothesis H2 tests if accruals quality (AQ) can determine the value relevance of earnings (VR). The results show that the higher the accruals quality, the higher the value relevance of earnings ($\beta = 0.2784$; $p < 0.05$).

These results confirm the predicted sign of the relationship between accruals quality and value relevance. The market prices accruals positively since they are explained by cash flows in the prior, current, and future periods. This finding suggests that accruals do not represent managerial choices but they reflect economic fundamentals. Earnings are superior to cash flows for making decisions because accruals are related to the kind and timing of the companies' operations, more than to the managerial discretion.

Finally, we test hypothesis H3 to find out if earnings smoothing (ESM) explains the value relevance of earnings (VR). The results indicate the existence of a positive relationship between them ($\beta = 0.4517$; $p < 0.01$).

This relationship means that higher levels of earnings smoothing determine higher value relevance of earnings. Therefore, we may conclude that, in this case, earnings smoothing can be considered as a manipulative practice that is informative for investors, as the results show that the more earnings are smoothed, the better they predict the market value of firms.

In summary, the results are consistent with managers' use of discretionary accruals to develop opportunistic behaviors, instead of to communicate information regarding future profitability. This is true except when managerial discretion is used to reduce the volatility of earnings, which improves the persistence and predictability of reported earnings. In this case, earnings management increases the informative value of earnings.

Table 2 includes the R-squared of the value relevance as the dependent variable in the model. This coefficient captures the variance of the dependent variable that is explained by the constructs that predict them. In this case, the model is able to explain more than 20 % of the variance of value relevance. 0.1 is established as the low value that a correct model should present.

Table 2. Results of structural model. R-squared of constructs

	R-square
VR	0.2089

5   Conclusions

This study investigates the impact of earnings quality on the relationship between earnings and stock prices. The empirical analysis evaluates if the earnings quality affects the stock pricing. Investors follow earnings as an important variable for decision-making and, in consequence, the quality of earnings should influence the relationship between earnings and stock prices. Specifically, we try to analyze the association between different desirable features of earnings and the investors' decision, using the relationship between earnings and stock prices.

In order to incorporate the multidimensional nature of earnings quality in our analysis, without the measurement of earnings quality being limited to one specific concept, we use Structural Equation Models. This methodology is suitable for the study of aspects related to earnings quality because, as we have seen, there are different measures that in turn are not directly observable variables (latent variables) and it allows us to base the study on different models to measure a same variable, in order to mitigate the effects of choosing the measurement method.

The results show that managers use discretionary accruals to develop opportunistic behaviors instead of to communicate information regarding future profitability. This is true except when managerial discretion is used to reduce the volatility of earnings, which improves the persistence and predictability of reported earnings. In this case, earnings management increases the informative value of earnings. Therefore, these results indicate that investors value the quality of accruals and dislike earnings manipulated by discretionary accruals except when this manipulation reduces the earnings volatility.

This paper contributes to the literature by evaluating how investors value various desirable features of earnings. This study also suggests an alternative approach to evaluate whether earnings management is opportunistic or informative. Apart from that, it provides a major methodological contribution to the study of earnings quality using structural equation models and an important compilation of the varied methods used in previous literature to measure each of the desirable earnings attributes related to earnings management.

References

1. Schipper, K., Vincent, L.: Earnings quality. Account. Horiz. **17**(Supplement), 97–111 (2003)
2. Francis, J., LaFond, R., Olsson, P., Schipper, K.: Costs of equity and earnings attributes. Account. Rev. **79**(4), 967–1010 (2004)
3. Bhattacharya, U., Daouk, H., Welker, M.: The world price of earnings opacity. Account. Rev. **78**(3), 641–678 (2003)
4. Subramanyam, K.: The pricing of discretionary accruals. J. Account. Econ. **22**(1–3), 249–281 (1996)
5. Chung, R., Ho, S., Kim, J.: Ownership structure and the pricing of discretionary accruals in Japan. J. Int. Account. Audit. Tax. **13**(1), 1–20 (2004)
6. Christensen, T.E., Hoyt, R.E., Paterson, J.S.: Ex Ante incentives for earnings management and the informativeness of earnings. J. Bus. Financ. Account. **26**(7–8), 807–832 (1999)
7. Marquardt, C., Wiedman, C.: The effect of earnings management on the value relevance of accounting information. J. Bus. Financ. Account. **31**(3–4), 297–332 (2004)
8. Dechow, P., Ge, W., Schrand, C.: Understanding earnings quality: a review of the proxies, their determinants and their consequences. J. Account. Econ. **50**(2–3), 127–466 (2010)
9. Gaio, C.: The relative importance of firm and country characteristics for earnings quality around the world. Eur. Account. Rev. **19**(4), 693–738 (2010)
10. Laksmana, I., Yang, Y.: Corporate citizenship and earnings attributes. Adv. Account. **25**(1), 40–48 (2009)
11. Dechow, P.M., Sloan, R.G., Sweeney, A.P.: Causes and consequences of earnings manipulation: an analysis of firms subject to enforcement actions by the SEC. Contemp. Account. Res. **13**(1), 1–36 (1996)
12. Burgstahler, D., Dichev, I.: Earnings management to avoid earnings decreases and losses. J. Account. Econ. **24**(1), 99–126 (1997)
13. García, O., Gill, B., de Albornoz, B., Gisbert, A.: La investigaciónsobre earnings managements. Span. J. Financ. Account. **34**(127), 1001–1033 (2005)
14. Healy, P., Palepu, K.: The effect of firms' financial disclosure policies on stock prices. Account. Horiz. **7**(1), 1–11 (1993)
15. Chaney, P., Jeter, D., Lewis, C.: The use of accruals in income smoothing: a permanent earnings hypothesis. Adv. Quant. Anal. Financ. Account. **6**(1), 103–135 (1998)
16. Demski, J.: Performance measure manipulation. Contemp. Account. Res. **15**(3), 261–286 (1998)

17. Arya, A., Glover, J.C., Sunder, S.: Are unmanaged earnings always better for shareholders? Account. Horiz. **17**(Supplement), 111–116 (2003)
18. Jiraporn, P., Miller, G.A., Yoon, S.S., Kim, S.: Is earnings management opportunistic or beneficial? An agency theory perspective. Int. Rev. Financ. Anal. **17**(3), 622–634 (2008)
19. Harris, T., Huh, E., Fairfield, P.: Gauging profitability on the road to valuation. Strategy Report, Global Valuation and Accounting. Morgan Stanley Dean Witter (2000)
20. Dechow, P., Dichev, I.: The quality of accruals and earnings: the role of accrual estimation errors. Account. Rev. **77**(Supplement), 35–59 (2002)
21. McNichols, M.: Discussion of the quality of accruals and earnings: the role of accrual estimation errors. Account. Rev. **77**(Supplement), 61–69 (2002)
22. Beidleman, C.: Income smoothing: the role of management. Account. Rev. **48**(4), 653–667 (1973)
23. Burns, N., Kedia, S.: The impact of performance-based compensation on misreporting. J. Financ. Econ. **79**(1), 35–67 (2006)
24. Bergstresser, D., Philippon, T.: CEO incentives and earnings management. J. Financ. Econ. **80**(3), 511–529 (2006)
25. Ronen, J., Sadan, S.: Smoothing Income Numbers: Objectives, Means, and Implications. Addison-Wesley, Reading (1981)
26. Healy, P., Wahlen, J.: A review of the earnings management literature and its implications for standard setting. Account. Horiz. **13**(4), 365–383 (1999)
27. Lambert, R.: Income smoothing as rational equilibrium behaviour. Account. Rev. **59**(4), 604–618 (1984)
28. Trueman, B., Titman, S.: An explanation for accounting income smoothing. J. Account. Res. **26**(Supplement), 127–139 (1988)
29. Sankar, M., Subramanyam, K.: Reporting discretion and private information communication through earnings. J. Account. Res. **39**(2), 365–386 (2001)
30. Dechow, P., Skinner, D.: Earnings management: reconciling the views of accounting academics, practitioners, and regulators. Account. Horiz. **14**(2), 235–250 (2000)
31. Leuz, C., Nanda, D., Wysocki, P.: Earnings management and investor protection: an international comparison. J. Financ. Econ. **69**(3), 505–527 (2003)
32. Tucker, J., Zarowin, P.: Does income smoothing improve earnings informativeness? Account. Rev. **81**(1), 251–270 (2006)
33. Ferrer, C.: Financial Information quality and its relevance for the stock market: measurement through earnings attributes, Doctoral Thesis, University of Zaragoza (2011)
34. Ferrer, C., Callao, S., Jarne, J.I., Laínez, J.A.: How do investors value earnings quality? An analysis with structural equation models. In: IX Workshop on Empirical Research in Financial Accounting (2012)

Evolving Improved Neural Network Classifiers for Bankruptcy Prediction by Hybridization of Nature Inspired Algorithms

Vasile Georgescu[✉] and Florentina Mihaela Apipie

Department of Statistics and Informatics,
University of Craiova, Craiova, Romania
v_geo@yahoo.com

Abstract. Bankruptcy prediction is a hard classification problem, as data are high-dimensional, non-Gaussian, and exceptions are common. Nature inspired algorithms have proven successful in evolving better classifiers due to their fine balance between exploration and exploitation of a search space. This balance can be further refined by hybridization, which may provide a good interplay of exploration (identifying new promising regions in the search space to escape being trapped in local solutions) and exploitation (using the promising regions locally, to search for eventually reaching the global optimum). The aim of this paper is to compare the performance of two search heuristics - Particle Swarm Optimization (PSO) and Gravitational Search Algorithm (GSA) – when using alone, or synergically, as a hybrid method, for evolving Neural Network (NN) classifiers for bankruptcy prediction.

Keywords: Nature inspired algorithms · PSO · GSA · Hybridization · Evolved NN classifiers · Bankruptcy prediction

1 Introduction

Using hybridization for achieving a good balance between exploration and exploitation became more and more influential since it was first advocated in a paper of Eiben and Schippers [1]. The exploration capability of nature inspired algorithms allows finding the most interesting basins of attractions (by extending the search over the whole sample space). They may thus avoid trapping in local minima, but are not as fast as local-search techniques when it comes to exploitation, suffering from low convergence rate in the last stage of approaching the solution. It is these complementary strengths that inspired the development of numerous hybrid algorithms, in hoping that the hybrids perform better than the individual algorithms. Exploration is sometimes associated with diversification, whereas exploitation is associated with intensification. Usually, the balance between them is insured by favoring exploration at the beginning of the search (when it is desirable to have a high level of diversification) and favoring exploitation at the end (when the algorithm is close to the final solution and intensifying the local search is more suitable).

Neural Networks (NN) are well suited for both classification and function approximation, and their training method has been originally based on

© Springer International Publishing Switzerland 2016
R. León et al. (Eds.): MS 2016, LNBIP 254, pp. 40–50, 2016.
DOI: 10.1007/978-3-319-40506-3_5

Back-Propagation (BP) algorithm. The standard version of the algorithm looks for the minimum of the error function in the weight space using the method of gradient descent. The combination of weights which minimizes the error function is considered to be a solution of the learning problem. The main drawback of BP is its tendency to become trapped in local minima of the error function. The success and speed of training depends upon the initial parameter settings, such as architecture, initial weights and biases, learning rates, and others. The idea of evolving NN by evolutionary algorithms, as an alternative to BP, can be traced back to the late 1980s, when Genetic Algorithms (GA) have been used with two distinct purposes: architecture optimization and weight training. The rationales behind this idea was that GA perform a more global search than NN with BP. BP takes more time to reach the neighborhood of an optimal solution, but then reaches it more precisely. On the other hand, GA investigate the entire search space. Hence, they reach faster the region of optimal solutions, but have difficulties to localize the exact point. Since the initiation of this direction of research, many nature inspired search heuristics have been employed to evolve NN. Two such heuristics are investigated in this paper – PSO and GSA, respectively – and their performances are compared when using alone, or synergically, as a hybrid method. The experimental setup for this comparison is that of evolving neural network classifiers for bankruptcy prediction.

2 Search Heuristics and Their Hybridization

2.1 Particle Swarm Optimization (PSO)

PSO has been originally proposed by Kennedy and Eberhart [3]. It is behaviorally inspired and belongs to Evolutionary Computation, whose main purpose is the emergence of complex behaviors from simple rules. In the specific case of PSO, the strategy of searching the problem hyperspace for optimum was developed out of attempts to model the social behavior of bird flocking or fish schooling.

PSO consists of a swarm of particles. Each particle resides at a position in the search space. The fitness of each particle represents the quality of its position. Initially, the PSO algorithm chooses candidate solutions randomly within the search space. The particles fly over the search space with a certain velocity. The velocity (both direction and speed) of each particle is influenced by its own best position found so far and the best solution that was found so far by its neighbors. Eventually the swarm will converge to optimal positions.

Let $i \in \{1, \ldots, N\}$, $x_i \in \Re^n$ and $v_i \in \Re^n$ be a particle, its position and its velocity, respectively. Now, consider a fitness function $f : \Re^n \to \Re$. Candidate solutions x_i are initially placed at random positions in the search-space and moving in randomly defined directions. The direction of a particle is then gradually changed to move in the direction of the best found positions of itself and its peers, searching in their vicinity and potentially discovering better positions.

The pseudo-code of PSO is given below:

1. Initialize all particles i with random positions in the search space: $x_i^0 \sim U(b_{lo}, b_{up})$, where b_{lo} and b_{up} are the lower and upper boundaries of the search-space.

2. Set each particle's best known position to its initial position: $pBest_i^0 = x_i^0$.

3. Initialize each particle's velocity to random values: $v_i^0 \sim U(-d, d)$, where $d = |b_{up} - b_{lo}|$.

4. Set the initial swarm's best known position $gBest^0$ to the $pBest_i^0$ for which $f(pBest_i^0)$ is lowest.

5. **repeat**

6.   **for** all Particle i in the swarm do

7.   Pick two random numbers: ε_p, $\varepsilon_g \sim U(0, 1)$.

8.   Update the particle's velocity:

$$v_i^{t+1} = w \cdot v_i^t + c_p \cdot \varepsilon_p \cdot (pBest_i^t - x_i^t) + c_g \cdot \varepsilon_g \cdot (gBest^t - x_i^t). \qquad (1)$$

where w is a parameter, called inertia weigth, c_p is the so-called self adjustment coefficient, c_g is the so-called social adjustment coefficient, x_i^t is the current position of particle i at iteration t, $pBest_i^t$ is the best position in the current neighborhood, and $gBest$ is the best position so far.

9.   Compute the particle's new position:

$$x_i^{t+1} = x_i^t + v_i^{t+1}. \qquad (2)$$

10.   **if** $f(x_i^{t+1}) < f(pBest_i^t)$ **then**

11.   Update the particle's best known position:

$$pBest_i^{t+1} = (x_i^{t+1}). \qquad (3)$$

12.   **end if**

13.   **if** $f(pBest_i^{t+1}) < f(gBest^t)$ **then**

14.   Update the swarm's best known position:

$$gBest^{t+1} = pBest_i^{t+1}. \qquad (4)$$

15.   **end if**

16.   **end for**

17. **until** termination criterion is met

18. return the best known position: $gBest$.

The first term of (1), $w \cdot v_i^t$, is the inertia component, responsible for keeping the particle moving in the same direction it was originally heading. The role of the coefficient w is either to damp the particle's inertia or to accelerate the particle in its original direction. Generally, lower values of w speed up the convergence of the swarm to optima, and higher values of w encourage exploration of the entire search space.

2.2 Gravitational Search Algorithm (GSA)

Gravitational search algorithm (GSA) was originally proposed by Rashedi et al. [5]. In GSA, all particles are viewed as objects with masses. Based on the Newton's law of universal gravitation, the objects attract each other by the gravity force, and the force makes all of them move towards the ones with heavier masses. Each mass has four characteristics: position, inertial mass, active gravitational mass, and passive gravitational mass. The first one corresponds to a solution of the problem, while the other three are determined by fitness function.

Let us consider a system with N masses (agents), where the ith mass's position is defined as follows:

$$X_i = \left(x_i^1, \ldots, x_i^d, \ldots, x_i^n\right), \quad i = 1, 2, \ldots, N. \tag{5}$$

The gravitational force acting on mass i from mass j at a specific time t is defined as follows:

$$F_{ij}^d(t) = G(t) \frac{M_{pi}(t) \cdot M_{aj}(t)}{R_{ij}(t) + \varepsilon} \left(x_j^d(t) - x_i^d(t)\right), \tag{6}$$

where M_{aj} is the active gravitational mass related to agent j, M_{pi} is the passive gravitational mass related to agent i, $G(t) = G_0 \cdot e^{-\alpha \cdot iter/maxiter}$ is a gravitational constant that is diminishing with each iteration, ε is a small constant, and $R_{ij}(t) = \left\|X_i(t) - X_j(t)\right\|_2$ is the Euclidian distance between two agents i and j.

For the purpose of computing the acceleration of an agent i, total forces (related to each direction d at time t) can be defined by

$$F_i^d(t) = \sum_{j=1, j\neq i}^{N} \varepsilon_j F_{ij}^d(t), \quad \varepsilon_j \sim U(0, 1), \tag{7}$$

Alternatively, to improve the performance of GSA by controlling exploration and exploitation, only the group *Kbest* of heavier agents is allowed to attract the others, where *Kbest* is decreasing over time.

$$F_i^d(t) = \sum_{j \in Kbest, j\neq i}^{N} \varepsilon_j F_{ij}^d(t), \quad \varepsilon_j \sim U(0, 1). \tag{8}$$

Thus, by lapse of iterations, exploration is fading out and exploitation is fading in.

Given the inertial mass M_{ii} of the ith agent, we can now define the acceleration of the agent i, at time t, in the dth direction:

$$a_i^d = \frac{F_i^d(t)}{M_{ii}(t)}. \tag{9}$$

The ith agent's next velocity and position can then be computed as:

$$v_i^d(t+1) = \eta_i \cdot v_i^d(t) + a_i^d, \quad \eta_i \sim U(0,\ 1), \tag{10}$$

$$x_i^d(t+1) = x_i^d(t) + v_i^d(t+1). \tag{11}$$

Finally, after computing current population's fitness, the gravitational and inertial masses can be updated as follows:

$$m_i(t) = \frac{fit_i(t) - worst(t)}{best(t) - worst(t)}, \tag{12}$$

$$M_i(t) = \frac{m_i(t)}{\sum\limits_{j=1}^{N} m_j(t)}. \tag{13}$$

where $fit_i(t)$ is the fitness value of the agent i at time t; $best(t)$ is the strongest agent at time t, and $worst(t)$ is the weakest agent at time t; $best(t)$ and $worst(t)$ are calculated as:

$$\text{For a minimization problem}: \begin{cases} best(t) = \min\limits_{j \in \{1,\ ...,\ N\}} fit_j(t), \\ worst(t) = \max\limits_{j \in \{1,\ ...,\ N\}} fit_j(t). \end{cases} \tag{14}$$

$$\text{For a maximization problem}: \begin{cases} best(t) = \max\limits_{j \in \{1,\ ...,\ N\}} fit_j(t), \\ worst(t) = \min\limits_{j \in \{1,\ ...,\ N\}} fit_j(t). \end{cases} \tag{15}$$

The steps of implementing GSA can be summarized as follows:

1. Generate the initial population.
2. Evaluate the fitness for all agents.
3. Update the parameters $G(t)$, $best(t)$ and $worst(t)$.
4. Calculate the gravitational and inertial masses $m_i(t)$ and $M_i(t)$ and the total forces $F_i^d(t)$ in different directions, for $i = 1, 2, ..., N$.

5. Update the velocities v_i^d and the positions x_i^d.
6. Repeat steps 2 to 5 until the stop criterion is reached. If a specified termination criterion is satisfied, stop and return the best solution.

2.3   The PSOGSA Hybrid Algorithm

Hybridization itself is an evolutionary metaheuristic approach that mainly depends upon the role of the parameters in terms of controlling the exploration and exploitation capabilities. In principle, we can exploit synergically the mechanisms of control from two algorithms in order to form a hybrid with combined capabilities. This may be more likely to produce better algorithms. The critical parameters in PSO are *pBest*, whose role is to implement the exploration ability, and *gBest*, whose role is to implement the exploitation ability. The critical parameters in GSA are G_0 and α, which determine the values of $G(t)$, i.e., $G(t) = G_0 \cdot e^{-\alpha \cdot iter/maxiter}$. They allow the fine tuning of the exploitation capability in the first stage and a slow movement of the heavier agents in the last stage. Numerical experiments have shown that PSO performs better in exploitation, whereas GSA performs better in exploration. However, the latter suffers from slow searching speed in the last iterations. A new hybrid algorithm, called PSOGSA, has been developed by combining the mechanism of these two algorithms and the functionality of their parameters. It was recently proposed by Mirjalili et al. [4] and has been tested on twenty-three benchmark functions in order to prove its higher performance compared to standard PSO and GSA. The results shown that PSOGSA outperforms both PSO and GSA in most cases of function minimization and that its convergence speed is also faster.

The main difference in PSOGSA is the way of defining the equation for updating the velocity $V_i(t)$:

$$V_i(t+1) = w \cdot V_i(t) + c_1' \cdot \xi_1 \cdot ac_i(t) + c_2' \cdot \xi_2 \cdot (gBest - X_i(t)) \qquad (16)$$

where c_1' and c_2' are two adjustable parameters, w is a weighting coefficient, ξ_1, $\xi_2 \sim U(0, 1)$ are random numbers, $ac_i(t)$ is the acceleration of agent i at iteration t, and *gbest* is the best solution so far.

The way of updating the agent positions is unchanged:

$$X_i(t+1) = X_i(t) + V_i(t+1) \qquad (17)$$

By adjusting the parameters c_1' and c_2' via the updating procedure, PSOGSA has a better ability to balance the global search and local search. The agents near good solutions try to attract the other agents which are exploring the search space. When all agents are near a good solution, they move very slowly. By using a memory to store the best solution (*gBest*) found so far, PSOGSA can exploit this information, which is accessible anytime to each agent. Thus the agents can observe the best solution and can tend toward it.

3 Using PSO, GSA and PSOGSA to Evolve Neural Network Classifiers for Bankruptcy Prediction

3.1 NN Architecture, Fitness Function and the Encoded Strategy

The architecture of the evolved NN is determined by its topological structure and can be described as a directed graph in which each node performs a transfer function, typically a sigmoid:

$$f(s_j) = \frac{1}{1 + e^{-\left(\sum_{i=1}^{n} w_{ij} \cdot x_i - \theta_j\right)}}, \quad j = 1, 2, \ldots, h \tag{18}$$

where $s_j = \sum_{i=1}^{n} w_{ij} \cdot x_i - \theta_j$, n is the number of the input nodes, w_{ij} is the connection weight from the ith node in the input layer to the jth node in the hidden layer, θ_j is the bias (threshold) of the jth hidden node, and x_i is the ith input.

The final output of the NN can be defined as follows:

$$o_k = \sum_{j=1}^{h} w_{kj} \cdot f(s_j) - \theta_k, \quad k = 1, 2, \ldots, m, \tag{19}$$

where w_{kj} is the connection weight from the jth hidden node to the kth output node and θ_k is the bias (threshold) of the kth output node. The architecture of a NN with 2 inputs, 2 outputs and 3 hidden nodes is shown in Fig. 1.

Evolving NN with search heuristics consists of using that heuristic to find the parameters (weights and biases) of the NN as a solution of an optimization problem.

As an encoding strategy we use a connection-based direct encoding of the NN parameters, such as the weights and biases, which are passed, as candidate solutions, to the fitting (objective) function of the population-based optimization algorithm.

The fitness function is defined in terms of the Mean Square Error (MSE) of the NN. Let us denote by q the number of training samples, by d_i^k the desired output of the ith

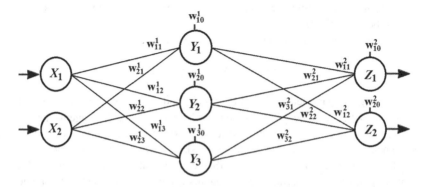

Fig. 1. The architecture of a NN with 2 inputs, 2 outputs and 3 hidden nodes.

input unit when the kth training sample is used, and by o_i^k the actual output of the ith input unit when the kth training sample is used. Then:

$$MSE = \sum_{k=1}^{q} \frac{E_k^2}{q}, \quad \text{where} \quad E_k^2 = \sum_{i=1}^{m} \left(o_i^k - d_i^k \right)^2. \tag{20}$$

3.2 The Experimental Setup

We next discuss the experimental setup proposed in this paper. Our goal is to compare the performances of PSO, GSA and PSOGSA, when using to evolve neural network classifiers for bankruptcy prediction.

A sample of 130 Romanian companies has been drawn from those listed on Bucharest Stock Exchange (BSE), with the additional restriction of having a turnover higher than one million EURO. Financial results for the selected companies were collected from the 2013 year-end balance sheet and profit and loss account, and were taken from the official website of BSE, www.bvb.ro. As predictors, a number of 16 financial ratios have been used in our models. The classification task consists of building classification models from a sample of labeled examples and is based on the search for an optimal decision rule which best discriminates between the groups in the sample.

When evaluating the predictive performance of one or more models, one of the core principles is that out-of-sample data is used to test the accuracy. The validation method we used in our experiments is the holdout method. The data set has initially been split into two subsets; about 60 % of the data have been used for training and 40 % for testing.

In binary classification, the accuracy is a statistical measure of how well a classifier correctly identifies if an object belongs to one of two groups. However, accuracy is not a reliable metric for the real performance of a classifier, because it will yield misleading results if the data set is unbalanced. Unfortunately, this is the case with our dataset, where the number of samples in the two classes varies greatly: 104 solvent firms and 26 insolvent ones. Thus, a more detailed analysis than mere proportion of correct guesses (accuracy) is needed. Actually, the performance of the competing classifiers was evaluated using the Confusion Matrix and the Receiver Operating Characteristic (ROC) analysis.

3.3 Results

The NN classifiers evolved with PSO, GSA and PSOGSA have similar architectures: 16 inputs, 2 outputs (binary response) and 15 hidden nodes. Our results validate the superiority of the hybrid PSOGSA algorithm that outperforms both the PSO and GSA. Figure 2 shows the learning performance of using PSOGSA to evolve the NN. Figures 3 and 4 show the confusion matrices and ROC curves for training and test datasets.

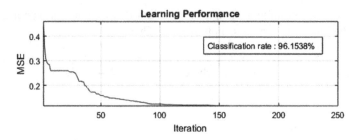

Fig. 2. NN evolved with PSOGSA: learning performance

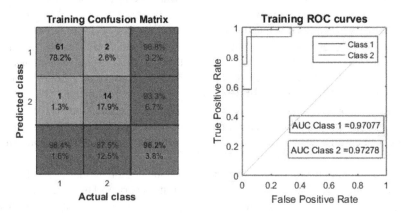

Fig. 3. NN evolved with PSOGSA: training confusion matrix and ROC curves (Color figure online)

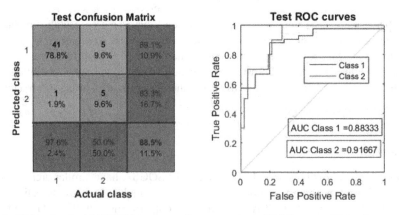

Fig. 4. NN evolved with PSOGSA: test confusion matrix and ROC curves (Color figure online)

Table 1. In-sample and out-of-sample average classification error rates

	PSOGSA	PSO	GSA
In-sample classification rate	96.2 %	92.3 %	93.6 %
Out-of-sample classification rate	88.5 %	86.5 %	86.5 %

The results for all search heuristics used to evolve NN classifiers for our application at hand (bankruptcy prediction) are summarized in Table 1.

As we expected, the in-sample classification rates are better than the out-of-sample classification rates for all three algorithms. However PSOGSA outperforms PSO and GSA in case of already seen, as well as unseen data. Further experiments are intended in order to evaluate repeatedly the performance of the three algorithms in terms of average classification rates, or to apply them on other databases.

Assessing the performance of the above search heuristics in cases when neural networks for function approximation (regression) are evolved is another way to be considered.

4   Conclusion

The aim of this paper was to compare the performances of two search heuristics – Particle Swarm Optimization (PSO) and Gravitational Search Algorithm (GSA) – when using alone, or synergically, as a hybrid method, for evolving Neural Network (NN) classifiers for the purpose of bankruptcy prediction. Hybridization has often proven to be a successful approach in many cases where, by combining the mechanism of two algorithms and the functionality of their parameters, we are able to find new ways of controlling the exploration and exploitation capabilities of the newly generated hybrid. The results reported here may be seen as another step in validating approaches of this kind.

References

1. Eiben, A.E., Schippers, C.A.: On evolutionary exploration and exploitation. Fundam. Inform. **35**(1–4), 35–50 (1998)
2. Georgescu, V.: Using genetic algorithms to evolve a type-2 fuzzy logic system for predicting bankruptcy. In: Gil-Aluja, J., Terceño-Gómez, A., Ferrer-Comalat, J.C., Merigó-Lindahl, J. M., Linares-Mustarós, S. (eds.) Scientific Methods for the Treatment of Uncertainty in Social Sciences. Advances in Intelligent Systems and Computing, vol. 377, pp. 359–369. Springer, Heidelberg (2015)
3. Kennedy, J., Eberhart, R.C.: Particle swarm optimization. In: Proceedings of IEEE International Conference on Neural Networks, vol. 4, pp. 1942–1948 (1995)
4. Mirjalili, S., Mohd Hashim, S.Z.: A new hybrid PSOGSA algorithm for function optimization. In: International Conference on Computer and Information Application (ICCIA 2010), pp. 374–377 (2010)

5. Rashedi, E., Nezamabadi, S., Saryazdi, S.: GSA: a gravitational search algorithm. Inf. Sci. **179** (13), 2232–2248 (2009)
6. Shi, Y., Eberhart, R.C.: A modified particle swarm optimiser. In: IEEE International Conference on Evolutionary Computation, Anchorage, Alaska (1998)

Environmental Performance Assessment in the Apparel Industry. A Materiality-Based Approach

Raúl León[1(✉)], Idoya Ferrero-Ferrero[2],
and María Jesús Muñoz-Torres[2]

[1] Accounting and Finance Department,
University of Zaragoza, Teruel, Spain
rleon@unizar.es
[2] Finance and Accounting Department,
University Jaume I, Castellón de la Plana, Spain
{ferrero,munoz}@uji.es

Abstract. Materiality has become a relevant issue in sustainability disclosure and business performance assessment. The main objective of this paper is to provide an assessment framework of sustainability performance based on expert knowledge. The expert knowledge has been extracted from the materiality analysis included in the sustainability reports. Focusing on a sectorial approach, this paper has applied the proposed methodology in a sample of 57 wearing apparel firms. The main results reveal that the most relevant aspects are 'Effluents and Waste', 'Emissions' and the consumption of resources ('Energy', 'Materials', and 'Water'), which simbolise almost the 60 % of the weights of the aspects. In addition, this study shows that those firms with a higher environmental performance represent a reduced percentage of the total sales of the sector and are, mainly, located in North America and Europe.

Keywords: Sustainability · Corporate social responsibility · Performance assessment · Materiality analysis · Wearing apparel industry

1 Introduction

During the last few years, research on social and environmental performance of firms has become a hotspot for practitioners and academia [1, 2]. In addition to financial performance, assessment and disclosure of social and environmental performances of corporations are considered issues of huge relevance for the different stakeholders of any company, since they strongly determine the survival of companies in the long term and increase their value [3, 4], and also provide insights on how a specific company manages its risks and contributes to sustainable development.

Given the increasing demand for greater transparency on both environmental and social performance, corporations are adopting sustainability reporting in their disclosure practices. Sustainability reports have become the main resource of information for stakeholders, and they constitute the main tool for assessing the social and environmental performance of corporations.

R. León et al. (Eds.): MS 2016, LNBIP 254, pp. 51–60, 2016.
DOI: 10.1007/978-3-319-40506-3_6

However, sustainability reporting remains voluntary and largely unregulated in most countries [5] and the current disclosure practices still have significant gaps which have come to light in numerous and recent surveys. For instance, EYGM conducted a survey of institutional investors on non-financial performance and it revealed that institutional investors never considered in their decision making process the environmental, social and governance (ESG) information that companies had published [6]. The explanation was that investors had a lack of information to understand what issues could materially impact on value creation and they often failed to be able to compare performance, even among peer organisations. In this regard, stakeholders require relevant, standardized and comparable information to make sustainable decision. In addition, several scholars note that sustainability reports show incomplete and irrelevant information for stakeholders, which is generally considered too generic and lacking of detailed quantifiable measures and comparable information [7, 8], and they highlight that companies tend to both over-report not material issues and under-report material issues [9]. In these circumstances, it is difficult to assess the performance of corporations according to what stakeholders consider important.

Therefore, one of the challenges of sustainability reporting is to identify the material aspects for both internal and external stakeholders. The concept of materiality is understood as those issues of greatest importance. In particular, Global Reporting Initiative (GRI) defines materiality as the degree of importance at which an aspect becomes sufficiently important to be reported. According to GRI, the threshold for defining material topics should be set to identify those opportunities and risks which are most important to stakeholders, the economy, environment and society [10]. In a similar way, other institutions such as the Sustainability Accounting Standards Boards link the materiality of an issue with the degree in that this issue is important to investors or other stakeholders in making their decisions. Therefore, one might expect that materiality, as a measure of importance between aspects, can be used in performance assessment to calculate the aggregation of the different aspects in a synthetic measure of performance.

Traditionally, the assessment of sustainability performance has been developed by academics and practitioners using synthetic indicators, which allow reducing the multidimensionality and simplifying the complexity of sustainability. However, the underlying dimensions of indices are not clearly defined and the opacity of the methodologies employed calls into question the reliability and comparability of scores [11]. In this regard, this study expects to contribute to literature in two directions. First, this study provides an assessment framework of sustainability performance based on expert knowledge that combines the simplicity required at corporate level in order to provide useful information and the scientific rigor to make the scores reliable and comparable. Second, this study digs deeper into the materiality assessment for defining report content, integrating the materiality analysis provided by companies in the assessment framework, which could be applied in future studies to test the robustness of the materiality analysis.

Different organisations such as the International Integrated Reporting Council (IIRC), the Sustainability Accounting Standards Boards (SASB) and the GRI have developed and up-dated reporting guidelines based on the materiality principle. However, each different reporting initiative involves a different subjective process to assess materiality, which could lead to selective reporting and a loss of credibility in sustainability reporting. This study is focused on the framework of GRI, since it is considered as the primary example of sustainability reporting and it has been widely applied in multi-national firms that operate in a variety of industries [12, 13].

Material aspects differ substantially across industries. Therefore, an approach focused on a limited number of the most relevant aspects for a specific industry can contribute to improve comparability and practicability for stakeholders. This study adopts an industry-based approach and attempts to advance the materiality framework in the wearing apparel industry. The wearing apparel industry is considered as one of the most unsustainable industries in the world from an environmental dimension [14]: use of harmful chemicals, high consumption of water and energy, generation of large quantities of solid and gaseous wastes, spillages, huge fuel consumption for transportation to remote places where textile units are located, and use of non-biodegradable packaging materials. Given the relevance of environmental impacts of this industry, this paper provides a first insight on the environmental dimension of sustainability in the wearing apparel sector.

The main objective of this paper is to develop a framework for environmental performance assessment based on a materiality analysis approach. To this end, this study uses different methodologies to provide an integrated assessment of corporations, calculated upon different performance indicators, which have been weighted according to the materiality of the aspects each indicator is related to. The framework has been applied to the wearing apparel sector, offering interesting results about the materiality of different aspects within the industry and the assessments obtained by firms belonging to this industry.

This work is divided into five sections. After this introduction this study presents the theoretical background. Section 3 includes information on the sample and the methodology used in the empirical analysis. Section 4 presents the results and the final section offers the discussion and main conclusion.

2 Theoretical Framework

Nowadays, there are a large number of environmental indicators (e.g. Ecological Footprint or Material Input Per unit Service and Cumulative Energy Demand) that have contributed to the assessment of impacts at micro and macro scale. However, they have been especially criticized regarding their possible application at corporate level, since they are not appropriate to identify weaknesses in the environmental management and guide corporate policy. In addition, there is a lack of knowledge on how the existing indicators may be used jointly to achieve relevant and comprehensive evaluations. Consequently, there is a need to advance in the field to combine the simplicity required at corporate level for tracking and reporting environmental data, and the scientific rigor and to make the scores reliable [15].

A growing number of scholars [16, 17] and agencies (Trucost, KLD, SAM) have developed synthetic indices to reduce the complexity of environmental performance a small number of dimensions, although the methodologies employed are not always clearly defined [11].

This study attempts to advance sustainability assessment literature applying a methodology for the weighting and aggregation of information based on expert knowledge in order to avoid an information loss problem. Traditionally, the techniques to generate synthetic indicators such us factor analysis [18], data envelopment analysis [19] or fuzzy inference [20], have not implemented any external weighting system that assumes greater importance for certain indicators. Focusing on assessing sustainability performance at corporate level, a possible source of expert knowledge could be the materiality analysis that companies present in the sustainability reports.

Materiality in financial reporting is defined by Financial Accounting Standards Board (FASB, Statement of Financial Accounting Concept No. 2) as: 'The magnitude of an omission or misstatement of accounting information that, in the light of surrounding circumstances, makes it probable that the judgment of a reasonable person relying on the information would have been changed or influenced by the omission or misstatement'. The main message of this definition is that materiality establishes the threshold between what is important and trivial, and recently this has been extrapolated to sustainability context to decide the issues and indicators to include, omit, and emphasize in sustainability reporting.

The assessment of corporations according to their sustainability performance can resemble a multicriteria decision problem. Multicriteria decision making (MCDM) provides techniques for comparing and ranking different alternatives in the basis of diverse criteria. Therefore, by means of MCDM, companies can be ranked according to various sustainability aspects, considering both the performance of the companies in relation to each aspect, and the relative importance—materiality—of each aspect as assessment criteria. MCDM can therefore be used to identify those companies with better sustainability performance, that is, those presenting the highest degrees of performance for all the material criteria. Literature provides several MCDM methods, and among them Analytical Hierarchy Process (AHP) [21] and the Technique for Order of Preference by Similarity to Ideal Solution (TOPSIS) [22] are two of the most used.

3 Method

This section presents the method used in this research. It starts by presenting the criteria used to select the sustainability reports employed to weight the environmental criteria, the sample of companies that has been assessed and the performance indicators used in the process. The assessment method is then explained in detail.

According to GRI, material aspects are those that reflect the organization's significant economic, environmental and social impacts; or substantively influence the assessments and decisions of stakeholders. GRI defines a taxonomy of economic, social and environmental aspects, covering all the subjects that are addressed by the GRI indicator. To the purpose of this research, the 12 environmental aspects proposed by GRI have been selected, and 3 additional aspects related to product responsibility, due to their close connection to product environmental management.

The information about the materiality of the selected aspects in the wearing apparel industry is harvested from sustainability reports. The sample consists of those GRI sustainability reports of apparel companies listed in GRI database for year 2013 that included a materiality matrix (i.e., in total 9 reports). Using the materiality matrices of these nine companies, the relative weights for the selected aspects are able to be calculated according to the process below described.

Finally, the calculated weights have been used to estimate the corporate environmental performance of a sample of firms. The sample consists of 57 firms belonging to the wearing apparel industry, which has been selected according to the information available in Asset4 database. For each company in the sample, a set performance indicators corresponding to year 2013 have been extracted. Those indicators have been selected due to their relation to abovementioned material aspects, so that for each aspect all the indicators available in Asset4 and related to the aspect have been selected.

Regarding the assessment method itself, our proposal integrates steps from AHP and TOPSIS methods. Each selected step has been partially adapted to the available information and the specific purposes of this research. AHP is used to calculate the relative materiality—weight—of the different aspects that GRI proposes in relation to environmental performance, as well as to calculate the overall assessment of the sample in each of those aspects. In second phase TOPSIS is applied to the previous results with the aim of ranking the sample of companies on the basis of their environmental performance.

According to AHP, expert's opinions are used to build pair wise comparison matrices of the assessment criteria. The expert knowledge has been sourced from the materiality matrices of companies in the sector. Each of the materiality matrices has been considered as being an "expert", and comparison of aspects has been done using the relative position of aspects in the matrix. To this end, we rank aspects within each matrix according to the distance to the origin, and then we define the comparison values by rating the relative importance of the aspects according to their position in the rank and based on the scale proposed by Saaty [21] (1 = *equally important*; 3 = *weakly important*; 5 = *strongly important*; 7 = very *strongly important*; 9 = *extremely important*). Aspects not included in the matrices are considered *equally important* between them and all the aspects in the matrix *extremely important* in relation to those not included. The consistency index and the consistency ratio are then calculated for each matrix, in order to test the consistency of the judgments extracted from materiality matrices.

The matrices representing the individual judgments of the companies in the sample are aggregated to create a collective judgment matrix. To this end, we use the Aggregation of Individual Judgments (AIJ) method, one of the methods that have been found to be the most useful in AHP group decision making judgments (AIJ) [23]. According to AIJ, individual judgement matrices are aggregated by means of geometric mean to obtain a collective judgement matrix.

Let $D = \{d_1, d_2, ..., d_m\}$ be the set of materiality matrices collected from sustainability reports, $C = \{c_1, c_2, ..., c_m\}$ the set of sustainability aspects used as assessment criteria, and $A^{(K)} = a_{ij}^{(k)}{}_{n*n}$ the judgment matrix constructed by the process descried above from the materiality matrix d_k $(k = 1,2,...,m)$. AIJ can be used to calculate a collective judgement matrix $A^{(c)} = \left(a_{ij}^{(c)} \right)_{n*n}$ where

$$a_{ij}^{(c)} = \sqrt[m]{\prod_{K=1}^{m} a_{ij}^{(K)}}$$

After assessing consistency of the collective matrix, the next step entails using a prioritization method to derive a collective priority vector. The prioritization method refers to the process of calculating a priority vector $w = (w_1,...,w_n)^{\mathrm{T}}$, where $w_i \leq 1$ and $\sum_{i=1}^{n} w_i = 1$ from a judgement matrix A. To this end we follow Saaty's proposal, which is based on the eigenvalue method [24]. Let λ be the principal eigenvalue of A, and e^T the unique positive eigenvector of A that is normalised. Then, the priority vector w can be obtained by solving the linear system:

$$\begin{cases} Aw = \lambda w \\ e^T w = 1 \end{cases}$$

As a result, we obtain a priority vector with the relative importance—or materiality— of each sustainability aspect considered as criteria for the assessing the performance of corporations. Using this priority vector and a set of performance indicators associated to the different sustainability aspects, the performance of companies operating in the same industry can be ranked using the TOPSIS method.

In order to provide a single assessment for each criterion, performance indicators associated to each criterion are integrated in a single and normalised measure—value from 0 to 1—to be used as input for TOPSIS method. To simplify the use of TOPSIS, all the synthetic indicators are generated to represent a benefit criterion, standing 0 for the worst performance and 1 for the best.

Let $S = \{s_1, s_2, ..., s_p\}$ be the companies to be ranked and $R = r_{ij_{p*n}}$ the decision matrix with the measures previously calculated for each company s_i $(i = 1,2,...,p)$ and criteria c_j $(j = 1,2,...,n)$. The TOPSIS method starts by calculating the normalized decision matrix $R^{(N)} = \left(r_{ij}^{(N)} \right)_{p*n}$

$$r_{ij}^{(N)} = \frac{r_{ij}}{\sqrt{\sum_{i=1}^{p} r_{ij}^2}}, \quad i = 1,2,...p; j = 1,2,...n$$

By using the priority vector w generated with AHP, we calculate the weighted normalised decision matrix $T = (t_{ij}) = \left(r_{ij}^{(N)} w_j \right)_{p*n}$, and then we determine the ideal positive candidate $T^+ = (t_1^+, t_2^+, ... t_n^+)$ and negative ideal candidate $T^- = (t_1^-, t_2^-, ... t_n^-)$, where $t_j^+ = \max \left(t_{ij}^+ \right)$ and $t_j^- = \min \left(t_{ij}^- \right)$, $i = 1,2,...,p$; $j = 1,2,...,n$.

The distance of each company in the sample from the ideal positive and negative candidates are then calculated as:

$$d_i^+ = \sqrt{\sum_{j=1}^{n} \left(t_{ij} - t_j^+ \right)^2}, \quad i = 1, 2, \ldots p$$

$$d_i^- = \sqrt{\sum_{j=1}^{n} \left(t_{ij} - t_j^- \right)^2}, \quad i = 1, 2, \ldots p$$

The next step involves calculating the relative closeness to the ideal candidate, the one which is the nearest to the positive ideal candidate and the farthest from the positive ideal candidate. A closeness coefficient is defined to determine the relative distance the ideal solution, and it is calculated as

$$CC_i = \frac{d_i^-}{d_i^+ + d_i^-}, \quad i = 1, 2, \ldots p$$

According to the latter expression, company s_i is closer to d_i^+ and farther from d_i^- as CC_i approaches to 1. Therefore, according the de descent order of CC, it is possible to define a ranking of all companies and compare their environmental performance calculated on the basis of the materiality of the selected criteria.

4 Results and Discussion

The assessment method has been applied to the samples previously described. In relation to the weights of criteria, some differences have been found between the priority vector calculated from individual judgement matrices, which raise the question about the effectiveness and comparability of sustainability reports where contents are defined on the basis of different material aspects.

The collective judgment matrix derived from individual matrices has proven to be consistent, presenting a consistency index of 0.93 % and a consistency ratio of 0.59 %. The relative materiality of the different aspects is shown in Table 1. Results are consistent with previous qualitative studies [25], being 'Effluent and Waste', 'Materials', 'Water', 'Energy', and 'Emissions' the five most relevant aspects. It is important to note that the weights of these five aspects represent almost the 60 %, and therefore they play a crucial role in the assessment of companies. On the other side, aspects such as "Product and Service Labeling", "Compliance" and "Environmental Grievance Mechanisms" seem to be almost immaterial for companies in the wearing apparel industry. These results should therefore be contrasted with other studies, since they may evidence a lack of suitability in the selection of material aspects GRI proposes in its guidelines for general application.

Table 1. Priority vector for material environment aspects

Aspect	Materiality
Effluents and waste	16,56 %
Materials	14,42 %
Water	10,62 %
Energy	9,65 %
Emissions	8,44 %
Customer health and safety	6,98 %
Supplier environmental assessment	5,50 %
Products and services	5,46 %
Biodiversity	4,71 %
Marketing communications	4,02 %
Overall	3,56 %
Transport	3,13 %
Product and service labeling	2,78 %
Compliance	2,34 %
Environmental grievance mechanisms	1,83 %

The obtained weights, which are included in Table 1, have been used to assess the environmental performance of the sample extracted from Asset4, which contains 57 companies in the wearing apparel industry. The sample reveals that the industry is quite concentrated, with 2 companies representing the 52 % of total revenue—and 18 the 95 %—being Japan the most relevant region. In general terms, the results show that bigger companies do not appear in the first positions of the environmental performance ranking. The first 22 companies in the ranking represent the 51 % of revenue. However, it is important to remark that the biggest company of the sample is ranked as 22nd, which represents 32.5 % of total revenue. In fact, the top ten in the ranking only represent 3.26 % of total sales, and these first positions are placed by companies from North America, Europe and one unique company from China. Besides, the worst positions of the environmental performance ranking correspond to those companies that are located predominantly in Taiwan, Japan and Hong Kong. The environmental performance ranking seeks to illustrate how the assessment framework developed in this study takes into account the results of the materiality matrices of the sector and how the proposed methodology can be applied in a set of companies to aggregate and compare environmental information.

5   Conclusions

Materiality has become a hot topic in sustainability reporting and performance assessment. Based on the literature review, this paper proposes a method aimed at assessing organizations sustainability performance that integrates the materiality issue. The method is based on the materiality analysis of GRI sustainability reports and it is applied to the assessment of environmental performance of corporations in the wearing apparel industry.

The presented assessment framework is flexible, so diverse methods can be applied in diverse steps in order to test the robustness, and it can also be used for assessing companies in other industries or using different criteria. The results obtained from this study are satisfactory, but they can still be improved. The lessons from the application to the environmental assessment of firms in the wearing apparel sector can be added to the knowledge base of the industry and contribute to the development of tools for sustainability assessment.

The proposed method has many advantages for sustainability assessment researchers and practitioners. First, the use of pairwise judgment matrices may help to derive priority vectors for any set of assessment criteria, and they can be used to extract the knowledge from diverse sources, such us experts or stakeholders' opinion, or a collection of materiality matrices as in this study. Second, by ranking companies according to their sustainability performance assessment, stakeholders can better compare between companies, and use that in their decisions as customers, workers or investors.

One of the limitations of the paper is that we have applied the method in a restricted scenario, limiting the assessment criteria to the set of environmental material aspects proposed by GRI, using a small sample of materiality matrices for deriving the relative materiality—weight—of each aspect, and assessing a sample of companies according to the sample and data available in the Asset4 database. Therefore, this method and its application are still subjects for further research.

Acknowledgements. The authors wish acknowledge the support received from P1•1B2013-31 and P1•1B2013-48 projects through the Universitat Jaume I and Sustainability and Corporate Social Responsibility Master Degree (UJI–UNED). This research is partially financed by Nils Science and Sustainability Programme (ES07), ABEL – Coordinated Mobility of Researchers (ABEL-CM-2014A), and supported by a grant from Iceland, Liechtenstein and Norway through the EEA Financial Mechanism. Operated by Universidad Complutense de Madrid.

References

1. Moore, S.B., Manring, S.L.: Strategy development in small and medium sized enterprises for sustainability and increased value creation. J. Cleaner Prod. **17**(2), 276–282 (2009)
2. Kiron, D., Kruschwitz, N., Haanaes, K., Reeves, M., Goh, E.: The innovation bottom line. MIT Sloan Manag. Rev. **54**(2), 69–73 (2013)
3. Crifo, P., Forget, V.D., Teyssier, S.: The price of environmental, social and governance practice disclosure: an experiment with professional private equity investors. J. Corp. Finance **30**, 168–194 (2015)
4. Mackey, A., Mackey, T.B., Barney, J.B.: Corporate social responsibility and firm performance: investor preferences and corporate strategies. Acad. Manag. Rev. **32**(3), 817–835 (2007)
5. Manetti, G.: The quality of stakeholder engagement in sustainability reporting: empirical evidence and critical points. Corp. Soc. Responsib. Environ. Manag. **18**(2), 110–122 (2011)
6. EYGM : Tomorrow's investment rules, global survey of institutional investors on non-financial performance (2014). https://www.eycom.ch/en/Publications/20140502-Tomorrows-investment-rules-a-global-survey/download. Accessed Mar 2016

7. Hess, D.: Social reporting and new governance regulation: the prospects of achieving corporate accountability through transparency. Bus. Ethics Q. **17**(3), 453–476 (2007)
8. Dubbink, W., Graafland, J., Van Liedekerke, L.: CSR, transparency and the role of intermediate organisations. J. Bus. Ethics **82**(2), 391–406 (2008)
9. Font, X., Guix, M., Bonilla-Priego, M.J.: Corporate social responsibility in cruising: using materiality analysis to create shared value. Tourism Manag. **53**, 175–186 (2016)
10. Global Reporting Initiative: G4 Guidelines Part 1: Reporting Principles and Standard Disclosures (2013). https://www.globalreporting.org/resourcelibrary/GRIG4-Part1-Reporting-Principles-and-Standard-Disclosures.pdf. Accessed Mar 2016
11. Delmas, M.A., Etzion, D., Nairn-Birch, N.: Triangulating environmental performance: what do corporate social responsibility ratings really capture? Acad. Manag. Perspect. **27**(3), 255–267 (2013)
12. Joseph, G.: Ambiguous but tethered: an accounting basis for sustainability reporting. Crit. Perspect. Acc. **23**(2), 93–106 (2012)
13. Mahoney, L.S., Thorne, L., Cecil, L., LaGore, W.: A research note on standalone corporate social responsibility reports: signaling or greenwashing? Crit. Perspect. Acc. **24**(4), 350–359 (2013)
14. Choudhury, A.R.: Environmental impacts of the textile industry and its assessment through life cycle assessment. In: Roadmap to Sustainable Textiles and Clothing, pp. 1–39. Springer, Singapore (2014)
15. Herva, M., Franco, A., Carrasco, E.F., Roca, E.: Review of corporate environmental indicators. J. Cleaner Prod. **19**(15), 1687–1699 (2011)
16. Xie, S., Hayase, K.: Corporate environmental performance evaluation: a measurement model and a new concept. Bus. Strategy Environ. **16**(2), 148–168 (2007)
17. Walls, J.L., Phan, P.H., Berrone, P.: Measuring environmental strategy: construct development, reliability, and validity. Bus. Soc. **50**(1), 71–115 (2011)
18. Salvati, L., Zitti, M.: Substitutability and weighting of ecological and economic indicators: exploring the importance of various components of a synthetic index. Ecol. Econ. **68**(4), 1093–1099 (2009)
19. Martínez, F., Domínguez, M., Fernández, P.: El análisis evolvente de datos en la construcción de indicadores sintéticos: una aplicación a las provincias españolas. Estudios de Economía Aplicada **23**(3), 753–771 (2005)
20. Munda, G., Nijkamp, P., Rietveld, P.: Fuzzy multigroup conflict resolution for environmental management. In: The Economics of Project Appraisal and the Environment, pp. 161–183. Edward Elgar, Cheltenham (1994)
21. Saaty, T.L.: The Analytic Hierarchy Process. McGraw-Hill International, New York (1980)
22. Hwang, C.L., Yoon, K.: Multiple Attribute Decision Making: Methods and Applications. Springer, Heidelberg (1980)
23. Dong, Y., Zhang, G., Hong, W.C., Xu, Y.: Consensus models for AHP group decision making under row geometric mean prioritization method. Decis. Support Syst. **49**, 281–289 (2010)
24. Saaty, T.L.: Decision-making with the AHP: why is the principal eigenvector necessary. Eur. J. Oper. Res. **145**, 85–91 (2003)
25. Ferrero-Ferrero, I., León-Soriano, R., Muñoz-Torres, M.J., Fernández-Izquierdo, M.A, Rivera-Lirio, J.M., Escrig-Olmedo, E.: Materiality assessment in sustainability reports: a qualitative analysis of environmental aspects in the apparel industry. In: Fourth International Conference on Multinational Enterprises and Sustainable Development, Lisbon (2015)

Mutual Funds Survival in Spain
Using Self-Organizing Maps

Antonio Terceño, Laura Fabregat-Aibar[✉],
M. Teresa Sorrosal-Forradellas, and M. Glòria Barberà-Mariné

Faculty of Business and Economics, Department of Business Management,
University Rovira i Virgili, Av. Universitat 1, 43204 Reus, Spain
{antonio.terceno, laura.fabregat, mariateresa.sorrosal,
gloria.barbera}@urv.cat

Abstract. This paper presents an empirical analysis of the survival of mutual funds in Spain. The methodology used, Self-Organizing Maps, allows us to group mutual funds in survivor and non-survivor groups. The similarities and differences between the main features in each group determine the variables that explain the capacity of survival of the analysed funds. The results indicate that the variables that affect the disappearance of mutual funds are size, age, performance, volatility and the fees charged.

Keywords: Self-organizing maps · Mutual funds · Non-survivors funds · Survivors funds

1 Introduction

This article is a preliminary study of the variables that affect disappearance of mutual funds in Spain. We discuss the extent to which variables such as size, age, performance and the fees charged by the funds, could affect capacity of funds survival, as well as the relation between these variables.

The literature related to survival capacity tends to use econometric models with two or three explanatory variables [1–3].

We use a different methodology, Self-Organizing Maps (SOM), which are artificial neural networks that allow to incorporate multiple variables. In this paper, SOM are used to cluster mutual funds based on the variables included and thus we could analyse how they affect funds survival. Moreover, a key advantage of these networks is their unsupervised learning process that means that is not necessary to define groups previously; rather they will be defined according to the similarities and differences between the values of all variables included in the study. It is also important to note that the use of several variables is a novel aspect of our paper because, as we have mentioned, most of the existing studies are based on only two or three variables.

The purpose of this paper is to determine which characteristics of mutual funds are related to their disappearance.

The paper is structured as follows. Section 2 discusses the determinants which have an impact on survival capacity, by reviewing the literature about this topic.

© Springer International Publishing Switzerland 2016
R. León et al. (Eds.): MS 2016, LNBIP 254, pp. 61–68, 2016.
DOI: 10.1007/978-3-319-40506-3_7

Section 3 describes the methodology and how the data are processed, and in Sect. 4 the empirical results are shown. Finally, the conclusions are presented.

2 Determinants of Mutual Funds Survivorship

In order to select the variables, we need to know which ones have been studied in the literature and, therefore, which ones have already been analysed in terms of their effect on funds survival.

Some studies consider that the fund size is a key factor, suggesting that smaller funds have a higher probability of disappearance [1, 2, 4–15].

Closely related to this variables is the fund flows [7, 11–14]. These studies find a considerable decrease in assets inflow of mutual funds in the period before the fund closes.

Another strand of the literature focuses on the age of a fund. Some studies find that the probability of disappearance is inversely related to the age of the fund, suggesting that younger funds clearly have a higher probability as well as older funds are less likely to disappear [1, 3, 4, 9–11, 16, 17].

Another variable that has aroused interest and has been a recurring theme in the literature on mutual funds is performance in earlier years. Numerous studies find empirical evidence that a poor past performance increases the probability of disappearance [1–7, 9–15, 18]. In conjunction with the analysis of performance, some of the studies that analyse the risk of the fund by measuring volatility are [4, 6, 10].

The expenses of the fund are also a key factor that affects the capacity of survival, considering that funds with higher expense ratios have a higher probability of disappearance [2, 4–6, 9, 11, 12].

3 Data and Methodology

The databases are provided by the Spanish National Securities Market Commission (CNMV) and Morningstar Direct.

We use Self-Organizing Maps with a total random sample of 300 mutual funds that were alive at the beginning of 2012, of which 150 were dead during that year. We decided to give to the non-survivors funds and to the survivors funds the same weight because the aim of our study is to understand which variables have greater impact on survival capacity.

We use the following variables: (i) fund size, measured as the fund assets; (ii) age, calculated as the difference between 2012 and the year of creation; (iii) performance, taking into account two values: the return of the last year and the annualized return of the three previous years, so we can observe the effect of time on the variable; (iv) Risk, which includes volatility over one and three years, so that the effect of time can be also observed; finally, (v) expenses are incorporated as the total expense ratio (TER), percentage of the total fees management over the mean assets.

Table 1. Definition of variables

Code	Variable	Definition
Var 1	Age	Number of years since the creation of the fund until 2012
Var 2	Size	Total net assets (TNA) to 31/12/2012 or the date of disappearance, measured in euros
Var 3	1-Year Return	Annual return obtained by the fund in 2012
Var 4	3-Year Annualized Return	Three years annualized return obtained by the fund in 2012
Var 5	1-Year Standard Deviation	Annual standard deviation calculated from quarterly returns
Var 6	3-Year Standard Deviation	Standard deviation in the three previous years calculated from quarterly returns
Var 7	Annual Expense Ratio	Annual expense ratio charged by the fund in 2012

Table 1 summarizes these variables.

Table 2 presents the correlation matrix. We find a strong correlation between variable 5 (1-year Standard Deviation) and 6 (3-year Standard Deviation) and, therefore, we decide to exclude variable 6 to avoid overweighting, which could distort the results. Besides, some studies such as [4] or [18] consider that the performance of non-survivors funds is more representative whether the period studied is close to the year of disappearance. Thus, excluding variable 6, we consider a total of six variables (Table 3).

Table 2. Correlation matrix

	Var 1	Var 2	Var 3	Var 4	Var 5	Var 6	Var 7
Var 1	1						
Var 2	0.1002	1					
Var 3	0.3400	0.2273	1				
Var 4	−0.1128	0.0831	0.1524	1			
Var 5	0.0075	−0.1061	−0.2023	−0.2934	1		
Var 6	−0.0316	−0.1078	−0.2440	−0.2897	0.9484	1	
Var 7	−0.3038	−0.2411	−0.6352	0.0593	0.2212	0.2693	1

Table 3. Final selected variables

Code	Variable
Var 1	Age
Var 2	Size
Var 3	1-Year Return
Var 4	3-Year Annualized Return
Var 5	1-Year Standard Deviation
Var 6	Annual Expense Ratio

The SOM network is implemented in Matlab using the toolbox developed by the Laboratory of Information and Computer Science in the Helsinki University of Technology.

4 Empirical Application

When the network is implemented, it generates an output map of 11×8 (11 rows $\times$ 8 columns), forming 6 groups.

Figure 1 shows the Kohonen map, where the corresponding patterns (funds) have been numbered and "YES" or "NO" indicate whether the fund has disappeared or not.

We observe that groups situated at the bottom (groups 1, 2 and 3) contain basically the non-survivors funds, whereas those that are situated at the top (groups 4, 5 and 6) mainly represent clusters of survivors funds.

To analyze the behaviour of each pattern through its position in the map, it is necessary to evaluate the value of all the variables in the corresponding area. Figure 2 shows the value of all the variables in the map and their respective scale of values represented by colours.

The blue colour indicates the minimum values of each variable, while the red indicates the highest ones.

However, the best expected value for some variables is clearly the opposite for others. For example, the best situation for the expenses is represented by a low value (blue areas in variable 6), whereas for performance is desirable a high level, therefore the best colour in the scale for this variable is the red one.

Table 4 summarizes all the information about the characteristics of each group and whether they belong to survivors funds (SF) or non-survivors funds (NSF). We have used linguistic labels according to the colour scale shown in Fig. 2. Thus, we use 'high' when the cells of the group are red, 'medium-high' if cells are orange, 'medium' for yellow cells, 'medium-low' for turquoise and, finally, we have labelled as qualitative measure 'low' when cells are coloured in blue.

In general, there is an evident relation between some variables and the disappearance of mutual funds. We find that small and young funds with poor performance at one year and high expense ratios are more likely to disappear.

Regarding performance (variable 3 and 4), we find that the 3-year annualized return does not explain the disappearance of mutual funds because of the great disparity of their value in groups which are characteristics of survivor funds. However, the 1-year return is an influential factor on capacity of funds survival.

Age (variable 1) and size (variable 2) play a relevant role in the capacity of funds survival, since the probability of disappearing is higher for funds that are small and young.

Regarding expenses (variable 6), we observe that non-survivors funds have higher fees, so it is a significant variable on funds survival.

Finally, volatility (variable 5) is a variable that affects disappearance but only if it is accompanied by a low 1-year return because funds in group 4 have a high volatility but this is compensated by high return and, moreover, they are also old funds. Therefore, funds with these variables are more likely to live.

NO152 NO159 NO267 NO269	NO223 NO271 NO278	NO154 NO168 NO219 NO228 NO276 NO283 NO295 NO297	NO151 NO153 NO208 NO270	NO155 NO212 NO217 NO290	YES6 NO164 NO165 NO169 NO222 NO228 NO246 NO294	NO172 NO181 NO249	YES52 NO187 NO171 NO186 NO221
NO171 NO210	NO209 NO214	NO226	NO207 NO224 NO225 NO275	NO173 NO227 NO237	NO158	NO180 NO204	NO182 NO185 NO188 NO194 NO199 NO200 NO201 NO202 NO203 NO204 NO205
NO229 NO279 NO280	NO220 NO272 NO287	NO215 NO231 NO266 NO274 NO298 NO299	NO211 NO216	NO232 NO233 NO234 NO292	NO176 NO178 NO179 NO293	NO160 NO163 NO184 NO246 NO250 NO253 NO295	NO197 NO198 NO205
NO170 NO244 NO245 NO268	NO259	YES96 YES103	YES24 YES54 YES67 YES148 NO218 NO273 NO300	NO174	NO175 NO193 NO217 NO254 NO255 NO256	NO182 NO262	YES10 YES11 YES12 YES13 NO151 NO189 NO197 NO257
NO167 NO180 NO191 NO196 NO236 NO239 NO242 NO243 NO251 NO258 NO261		NO240 NO241 NO288	YES40 YES72 NO289 NO294	YES27 NO235	YES124	YES100 YES121	
YES18 NO197	YES14 NO285 NO296	YES33 YES37 YES71 YES78 NO291	YES110 YES112 NO281	YES42 YES114	YES8 YES82 NO260 NO262	NO195	YES7 YES9 YES23 YES74 YES94 YES101 YES135 NO158 NO206
NO166 NO213 NO238 NO277 NO282	YES15	YES108	YES28 YES128	YES3 YES84	YES76 YES77 YES115 YES116 YES134	YES50 YES133 NO193	YES1 YES34
YES34 NO280	YES150	YES29 YES119	YES36 YES46 YES56 YES93 YES99 YES113 YES141	YES22 YES102 YES140	YES120 YES149	YES142	YES5 YES21 YES30 YES59 YES70 YES90 YES117
YES60 YES129	YES39 YES80	YES58 YES87 YES118	YES46 YES57 YES64 YES68		YES43 YES49 YES139 YES143	YES19 YES35 YES41 YES44 YES83 YES88 YES131	YES47 YES63
YES53	YES52 YES109 YES138	YES127	YES62 YES69	YES86 YES92 YES136	YES132 YES146		YES25 YES38 YES61 YES65 YES73
YES95 YES105 YES111 YES125 YES126	YES55 YES81 YES97 YES98	YES26 YES66 YES79 YES89 YES91	YES2 YES85 YES122 YES147	YES17 YES106 YES107 YES123 YES130 YES145	YES16 YES104 YES144	YES51	YES20 YES31 YES48 YES75 YES137

Fig. 1. Self-organizing map for Spanish funds (Color figure online)

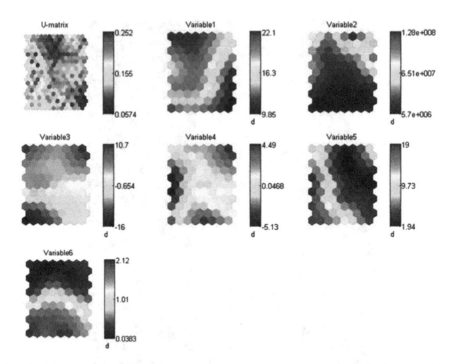

Fig. 2. Map of features (The scale of values next to each map shows the rank of values which are taken by the representative patterns of all the mutual funds located in one cell.) (Color figure online)

Table 4. Characteristics of each group

Group	Description	Age	Size	1-Year return	3-Year return	1-Year standard deviation	Annual expense ratio
1	NSF: 44	Low	Low	Low	Indefinite[1]	High	High
2	NSF: 49 SF: 6	Low	Low	Medium	Medium	Low	Medium-high
3	NSF: 35 SF: 2	Medium-high	Low	Medium	Medium	Low	Medium-high
4	NSF: 4 SF: 27	Medium-high	Low	Medium-high	Low	High	Medium. low
5	NSF: 9 SF: 61	High	Low	Medium-high	Medium	Medium-low	Low
6	NSF: 9 SF: 54	Medium-low	Medium-high	High	High	Low	Low

Note: when the variables have an indefinite value means that it is not possible to define as high, medium or low, since there are areas within the same group with different values for one variable.

As could be appreciate in Fig. 1, the groups are not totally homogeneous in relation to their disappearance. On the one hand, the survivors funds in groups 2 and 3, which are characteristic of disappeared funds, have made adjustments to its investment objectives, by modifying the initial conditions of mutual funds. This is the case, for example, of *Rural Mixto Internacional Flexible*, named NO260 in the map, or *Renta 4 Delta* denoted by NO195.

On the other hand, the disappeared funds in groups 4, 5 and 6, which are characteristic of survivors funds, belong to large fund families with a great number of funds with similar investment objectives. These funds are more likely to merge with other funds within the same family (*within-family merger*) because they have multiple funds with similar strategies and, consequently, need to combine redundant funds. This is the case of CaixaBank (with numbers YES10, YES11, YES12 and YES13). Moreover, we find several funds which have been necessarily merged because their companies have been absorbed by another one. This process is explained by the banking structuring that occurred in Spain during this period. This is the case of *Caixasabadell fondipòsit* (merger of BBVA with Unnim) which, on the map, is number YES148, *Bancaja Gestión Direccional 100* (merger of Bancaja with Bankia) with number YES15 and *Gesmadrid Renta Fija CP* (merger of Caja Madrid with Bankia) with number YES124.

5 Conclusions

In this paper we use Self-Organizing Maps to study how the variables age, size, performance, volatility and expenses affect the disappearance of Spanish mutual funds.

We obtain 6 groups according to the similarity between the variables included in the study.

Group 1 is only formed by non-survivors funds, which show the following behaviour of variables: high expenses, high volatility and low return and, moreover, these funds are small and young. Therefore, they are funds that tend to disappear.

Groups 2 and 3 are mainly formed by non-survivors funds, with a similar behaviour to the previous group. Even so, some of them survive; even though they have merged after the year of study or have modified their fund family.

Group 4 is related to survivors funds. Nevertheless, there is an interesting aspect to highlight. These funds have a low 3-year return though most of them remain alive because they have a high short-term return. So the effect of return in the short term is higher than in the long term, as long as funds are of a certain age and, therefore, are not unknown to investors.

Finally, groups 5 and 6 include funds with a good behaviour of the variables considered in this study. The only funds that disappear in these groups are for strategic reasons or for a desire to combine funds with similar strategies within the same fund family.

In summary, the results show that the analysed variables could explain the situation of each fund, positive or negative, except the 3-year return. In future research, we aim to include other variables and to examine whether Self-Organizing Maps can more accurately classify and define the characteristics of non-survivors funds.

References

1. Blake, D., Timmermann, A.: Mutual fund performance: evidence from the UK. Eur. Finan. Rev. **2**, 57–77 (1998)
2. Rohleder, M., Scholz, H., Wilkens, M.: Survivorship bias and mutual fund performance: relevance, significance, and methodical differences. Rev. Finan. **15**, 441–474 (2011)
3. Lunde, A., Timmermann, A., Blake, D.: The hazards of mutual fund underperformance: a cox regression analysis. J. Empirical Finance **6**, 121–152 (1999)
4. Brown, S., Goetzmann, W.: Performance persistence. J. Finan. **50**, 679–698 (1995)
5. Carhart, M.: On persistence in mutual fund performance. J. Finance **52**, 57–82 (1997)
6. Elton, E.J., Gruber, M.J.: Survivorship bias and mutual fund performance. Rev. Financ. Stud. **9**, 1097–1120 (1996)
7. Jayaraman, N., Khorana, A., Nelling, E.: An analysis of the determinants and shareholder wealth effects of mutual fund mergers. J. Finan. **57**, 1521–1551 (2002)
8. Carhart, M., Carpenter, J., Lynch, A., Musto, D.: Mutual fund survivorship. Rev. Financ. Stud. **15**, 1439–1463 (2002)
9. Cameron, A., Hall, A.: A survival analysis of Australian equity mutual funds. Aust. J. Manag. **28**, 209–226 (2003)
10. Zhao, X.: Exit decisions in the U.S. mutual fund industry. J. Bus. **78**, 1365–1402 (2005)
11. Ding, B.: Mutual fund mergers: a long-term analysis. http://ssrn.com/abstract=912927. (2006), Working Paper SSRN
12. Bris, A., Gulen, H., Kadiyala, P., Raghavendra, R.: Good stewards, cheap talkers, or family men? The impact of mutual fund closures on fund managers, flows, fees and performance. Rev. Financ. Stud. **20**, 953–982 (2007)
13. Andreu, L., Sarto, J.: Financial consequences of mutual fund mergers. Eur. J. Finan. 1–22 (2013)
14. Filip, D.: Survivorship bias and performance of mutual funds in hungary. Periodica Polytech. Soc. Manag. Sci. **22**, 47–56 (2014)
15. Boubakri, N., Karoui, A., Kooli, M.: Performance and survival of mutual fund mergers: evidence from frequent and infrequent acquirers. Working Paper (2012). doi:10.2139/ssrn. 2139430
16. Ter Horst, J., Nijman, T., Verbeek, M.: Eliminating look-ahead bias in evaluating persistence in mutual fund performance. J. Empirical Finan. **8**, 345–373 (2001)
17. Adkisson, J., Fraser, D.: Reading the stars: age bias in the morningstar ratings. Financ. Anal. J. **59**, 24–27 (2003)
18. Cogneau, P., Hübner, G.: The prediction of fund failure through performance diagnostics. J. Bank. Finance **50**, 224–241 (2015)

Modeling and Simulation in Business Management and Economy

An Estimation of the Individual Illiquidity Risk for the Elderly Spanish Population with Long-Term Care Needs

Estefania Alaminos[✉], Mercedes Ayuso, and Montserrat Guillen

Department of Econometrics, Riskcenter-IREA,
University of Barcelona, Av. Diagonal 690, 08034 Barcelona, Spain
{ealaminos,mayuso,mguillen}@ub.edu

Abstract. We study the individual illiquidity risk applied to the Spanish population aged over 64. This risk refers to the probability that an individual cannot afford long-term care costs and other expenses during their retirement. We analyse the case in which an individual is beneficiary of a contributory pension (a retirement or a widowhood pension) and has a degree of dependence (I, II, or III). Results show that illiquidity risk increases with age, and women present higher risk than men regardless of age and degree of dependence. The study reveals the vulnerable financial situation for individuals with moderate dependency, due to the low level of benefit that is received in this case.

Keywords: Dependency · Ageing · Risk quantification · Long-term care cost · Public benefits

1   Introduction

One of the main worries of pension savers is to be able to accumulate enough resources during active life. By enough resources we understand that income after retirement should be able to cover at least minimum expenses such as accommodation, subsistence, health and care services. Our work is motivated by the analysis of poverty in old-age. According to OECD [1] poverty rates are higher for older people than for the population as whole. A greater proportion of older women live in poverty than older men and old-age poverty rates increase with age. We understand that illiquidity refers to the situation where people have not enough cash to afford the costs that are incurred during their retirement. Estimation of illiquidity risk is relevant, especially after retirement, because it informs about the possibility of having an economic shortfall and so, identifying segments of the population with a high illiquidity risk reveals subgroups that demand intensive social services. They are especially vulnerable and may boost poverty.

From the individual perspective, a source of uncertainty for elderly people is when and to what extend they would need care or support. The probability of a person asking for long-term care services increases as age increases [2, 3]. When dependency arises, the different public benefits that are received are generally not enough to cover all expenses. In Spain, besides pensions (mainly retirement pensions as well as

© Springer International Publishing Switzerland 2016
R. León et al. (Eds.): MS 2016, LNBIP 254, pp. 71–81, 2016.
DOI: 10.1007/978-3-319-40506-3_8

widowhood pensions), people that have some degree of dependence may receive public benefits for long-term care if they have lost autonomy to perform daily activities such as being able to take care of themselves. However, there are levels of care need, especially for moderate cases, where public benefits are minimal or even non-existent, and therefore people have to face the costs by themselves or look for support among their relatives. Something similar may occur with more severe levels of dependency, where despite the existing public coverage, a pension income plus a long-term care benefit may not be sufficient to totally cover all care costs. Some authors indicate that most families do not have enough wealth to be able to afford the lifetime cost of care of the elderly members [4].

In Spain, long-term care protection has traditionally been provided by the family. Typically, women have borne the care of aged relatives [5]. Nevertheless, the progressive incorporation of women to the work force, low birth rates, and the rising longevity (also in dependence states [6]) have brought into question the sustainability of the long-term care public system [7] and the convenience of planning, before retirement, the capital necessities that may be required later on. In addition, reforms in public pension schemes have been carried out in recent years to guarantee their sustainability, which probably result in the short term in a reduction of the pension replacement rates, i.e. the percentage of the retirement pension amount with respect to the last wage perceived by the individual [8–10].

The retirement pension is the main source of income in Spanish households whose head of household is retired [11] and, therefore, pensions are their main source of liquidity. If costs exceed the pension amount, then families face illiquidity. Unlike other neighbouring countries, Spanish citizens generally make a substantial investment in their main residence during their working life [12], so that real states are the main savings instrument for elderly people [13], but they are well known to be illiquid. According to data extracted from the Spanish Survey of Household Finances [14], in households whose head is retired, real assets account for 82.7 % of total savings from which the main housing accounts for 62.0 %. In order to assess the financial resources of those who become dependent, we need to consider public benefits that they receive in addition to the contributory public pensions (mainly, retirement and widowhood). Benefits are granted whenever a person presents a significant level of severity of dependency and has been assessed to obtain public coverage.

The Spanish Law of Dependence[1] establishes three severity levels regarding long-term care needs that an individual may require: Degree 1 (moderate dependence), if support is needed at least once a day; Degree 2 (severe dependence), if assistance is to be provided two or three times per day, and Degree 3 (high dependence), if the individual needs assistance several times per day. Spanish legislation[2] sets the maximum amount of benefit per month that an individual may receive in concept of public benefits, which varies depending on the degree of dependence that has been recognised. Nevertheless, despite the fact that the Law is valid since 2006, benefits have been

---

[1] Law 39/2006, of December 14th.
[2] Royal Decree 1051/2013, of December 27th and Royal Decree-Law 20/12, of July 13th.

granted progressively beginning with the higher levels of severity. The coverage of moderate dependence is still not being granted in different parts of Spain.

The aim of this paper is to assess the illiquidity risk that people aged over 64 years might suffer in the case of needing long-term care. We quantify if a person who suffers dependence can afford expenditures related to this situation. The analysis is performed by each gender separately. The results show that illiquidity risk is not constant but increasing as age increases, especially among females. Usually, women with a retirement pension receive a lower amount than men due to shorter and discontinuous working periods [15]. Moreover, in general plenty of women only receive a low amount due to the nature of widowhood pensions [16]. In order to study this phenomenon, we have to consider prevalence rates of dependence for the elderly population by age groups and gender. The probability of being dependent increases with age, and is higher for women than for men [17].

The theoretical background or state of art about the topic and the method can be found in [2], where the prevalence rates from the EDAD survey [17] are presented. The methodology is developed according traditional disability actuarial models used in the literature [18–20].

The structure of this paper is as follows. In Sect. 2, we present a methodology to measure the illiquidity risk from a theoretical point of view, assessing the probability that the financial liquidity may turn negative depending on the age and the different dependence states. In Sect. 3, we present the data used in our study, as well as a descriptive analysis. In Sect. 4, we show the main results and describe different scenarios depending on the income received in terms of public pension (retirement or widowhood pension) and dependence subsidies. Finally, in Sect. 5, we summarize the main conclusions of this analysis.

2 Methodology

Considering the population aged over 64 years, we establish three different levels of dependence or severity, $h = \{a, b, c\}$, which depend on the intensity of long-term care needed. Thus, an individual can remain active (State 0), can be dependent with degree of severity a (State a), can be dependent with degree of severity b (State b), or dependent with degree of severity c (State c).

Depending on the severity level, individuals receive public benefits denoted by I_h, and need to afford long-term care costs G_h, therefore net expenses due to long-term care denoted by R_h is the result of the benefit received from the long-term care system, $R_h = I_h - C_h$. In this analysis, we consider that an individual is retired and receives a public pension P and he has basic living expenses G.

The main objective is to estimate illiquidity risk –denoted by r– by age and gender, this means, the probability that an individual, given a public retirement pension and a public dependence coverage (which is in accordance with his dependency level of severity), will be unable to afford basic living expenses G, and the long-term care costs C_h. The expenses' axis depending on the individual severity level of dependence is shown in Fig. 1.

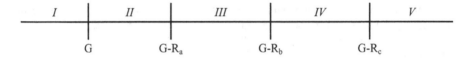

Fig. 1. Individual expenses' axis

We denoted by N_x the net amount available to a person aged x –defined in (1) –, than can be found from the public benefits that he receives, i.e. retirement pension and dependence benefit, and the expenses that he has to face:

$$N_x = P_x - G + I_h - C_h, \quad x = 65, \ldots, 100, \ h = a, b, c \tag{1}$$

where P_x is the public pension that the individual aged x receives, G are the basic living expenses that we establish constant for all ages, I_h is the public benefit due to dependency that is linked to the level of severity h, and C_h is the cost of long-term care depending on the degree of severity h. Note that in case of individual with no need of support, terms that are associated to long-term care costs and benefits $-C_h$ and I_h- are equal to zero. Letters I to V in Fig. 1 indicate the different scenarios that we consider for an individual as a function of their net amount N_x.

For instance, in case an individual has a net amount located in interval I, he cannot afford costs equal or over G, but if an individual has a net amount located in interval II, he can afford costs equal to G, but he cannot afford costs greater than $G - R_a$, meaning that he cannot afford the long-term care cost for levels of severity of dependence equal to a, b or c. In scenario III the individual can afford a cost between $G - R_a$, and $G - R_b$, then he cannot afford costs related to dependence level b or c, but he can afford severity level a. In scenario IV, individuals are able to afford costs related to dependence severity levels equal to a and b but not those related to severity level equal to c. Finally, in scenario V, individuals have net amount N_x which is larger than care cost plus subsistence, so they can afford any cost related to dependence regardless of the severity level.

The *illiquidity risk* for a person aged x is defined as the probability of his net income amount is less than zero:

$$r_x = \mathrm{Prob}(N_x < 0) = P(N_x < 0 | State_x = 0) \cdot P(State_x = 0)$$
$$+ P(N_x < 0 | State_x = a) \cdot P(State_x = a) + P(N_x < 0 | State_x = b) \cdot P(State_x = b) \tag{2}$$
$$+ P(N_x < 0 | State_x = c) \cdot P(State_x = c).$$

where probabilities related to being in a state are the probability of not being dependent (state 0) or being dependent with a certain severity level (a, b or c). In this work, the probability of being dependent with different degrees at different ages and by gender are estimated from the National Spanish Dependence Survey EDAD, conducted by the Spanish Statistics Institute in 2008. We denoted t_x^h as the probability of being dependent with the three different levels. Thus $1 - \sum_{h=a}^{c} t_x^h$ indicates the probability of not being dependent. Finally, *illiquidity risk* is calculated as follows:

$$r_x = \text{Pr}\,ob(N_x < 0) = P(N_x < 0 | State_x = 0) \cdot (1 - \sum_{h=a}^{c} t_x^h) + P(N_x < 0 | State_x = a) \cdot t_x^a$$

$$+ P(N_x < 0 | State_x = b) \cdot t_x^b + P(N_x < 0 | State_x = c) \cdot t_x^c.$$

(3)

A Monte Carlo simulation was carried out following the statistical distribution functions for the variables in (3) in order to obtain an estimate of the statistical distribution of the net amount N_x. An empirical quantile from the simulated distribution was calculated to obtain an estimate of the illiquidity risk r_x.

3 Data

In Spain, public benefits represent 72.0 % of the regular income of elderly people aged over 64 as reported by OECD data [21], this is significantly over the OECD average, which is about 59.0 %. On the other hand, 98.2 % of the beneficiaries of public benefits in Spain receive at least a pension from the Social Security System,[3] and the retirement pension is the principal source. Table 1 presents the number of pensioners. It is shown that men are the main beneficiaries of a retirement pension. A retirement average pension for a man aged over 64 is €1,135.51 per month. There are fewer women than men among beneficiaries of retirement pensions (1,932,566 women versus 3,254,971 men), and their average pension is much lower (€689.58 per month for women versus €1,135.51 per month for men). Furthermore, many women are only beneficiaries of a widowhood pension, with an average pension of roughly €625 per month.

In cases where a person is dependent and requires long-term care, the maximum amount per month established by law in Spain (Royal Decree 1051/2013, of December 27th; and Royal Decree-law 20/2012, of July 13th) to be received depending on the assessed degree of severity is shown in Table 2. The estimation of individual long-term

Table 1. Minimum, average and maximum by pension and number of pensioners. Contributory public retirement and widowhood pension data.

	Males		Females	
	Retirement	Widowhood	Retirement	Widowhood
Minimum amount	632.90	632.90	632.90	632.90
Average pension[a]	1,135.51	422.61	689.58	624.29
Upper limit	2,554.49	2,554.49	2,554.49	2,554.49
Number of pensioners	3,254,971	44,878	1,932,566	1,368,513

Pensions are expressed in euros.
Source: Spanish Ministry of Social Security, 2014. The average pension is the weighted average pension of the beneficiaries aged 65 and over. The number of pensioners refers to the retirement or widowhood pensioners, aged 65 and over.
[a] Note that in cases where the average pension does not reach the minimum amount established by law, the beneficiary receives an additional amount to achieve this minimum (those are called "minimum supplements").

---

[3] Spanish Social Security's statistics, 2014.

Table 2. Long-term care (LTC) benefits by level of severity, expressed in benefits per month, in Spain.

Severity level	Benefits
Degree I	300.00
Degree II	426.12
Degree III	715.07

Benefits are expressed in euros.
Source: Royal Decree-law 20/2012.

Table 3. Individual long-term care cost (LTC) depending on the level of dependence severity.

Severity level	LTC services	Individual annual cost, 2014
Degree I	3 h per day home assistance	15,111.00
Degree II	Day care centre and 1 h a day of home assistance	13,424.32
Degree III	Residence	18,084.85

Costs are expressed in euros.
Source: Authors' compilation from Spanish Ministry of Health, Social Services and Equality data, 2014. The cost of an hour of home care by a professional is set equal to €13.8 (31/12/2013); the annual cost in a day care centre is €8,387.32 (31/12/2012); and the annual cost of a residence is set equal to €18,084.85 (31/12/2012).

Table 4. Prevalence of dependence rates by age and gender, Spanish population.

Age	Males			Females		
	Degree I	Degree II	Degree III	Degree I	Degree II	Degree III
65	2.51 %	1.16 %	1.06 %	3.50 %	1.82 %	1.32 %
70	2.92 %	1.46 %	1.38 %	4.65 %	2.43 %	1.95 %
75	3.94 %	2.18 %	2.12 %	6.64 %	3.82 %	3.47 %
80	5.69 %	3.57 %	3.87 %	9.28 %	6.25 %	6.75 %
85	7.85 %	5.89 %	7.85 %	11.93 %	9.53 %	12.61 %
90	9.75 %	9.46 %	15.47 %	13.39 %	13.15 %	20.87 %
95	10.35 %	14.22 %	27.59 %	12.90 %	16.98 %	28.35 %

Source: [2].

care costs is shown in Table 3.[4] Prevalence rates of dependence of the Spanish population, by age and gender, depending on the degree of severity and estimated from de EDAD Survey [17] are shown in Table 4. These rates have been introduced in a

---

[4] The assistance needs regarding long-term care depend on the level of dependence (Table 3, column 2) and have been set based on the recommendations of expert geriatric groups, and have been used in several studies (among them, [2, 22–24]). These costs have been updated with the last data published by the Ministry of Health, Social Services and Equality, to each considered services.

previous work [2], and they reveal that prevalence increases with age regardless of the degree of severity. Furthermore, women present higher prevalence rates than men in any state of dependency.

Finally, as an indicator of basic consumption needs, we have used the Spanish Public Income Indicator of Multiple Effects (IPREM), which has remained equal to €532.51 per month during the last 7 years.

4   Results

Calculations have been carried out for Spanish population aged over 64 by gender. According to the methodology, we have three different degrees of severity, a, b and c. Depending on the degree of dependence recognised to the individual, he or she will receive an economic benefit (Table 2), where the annual amounts for entitlements are I_I = €3,600, I_{II} = €5,113.44, and I_{III} = €8,580.84, respectively. Annual long-term care costs regarding the scenarios considered in Table 3, are G_I = €15,111.00, G_{II} = €13,424.32 and G_{III} = €18,084.85. Then, based on our calculations, it is expected that the long-term care costs incurred by an individual are equal to R_I = €11,511, R_{II} = €8,310.88 and R_{III} = €9,504.01, respectively. This means that these costs are not covered by the dependence benefit that the individual receives from the public long-term care system. Thus, the annual total costs that the individual may face taking into account the basic expenses (which we have supposed as an amount of money equal to the annual IPREM €6,390.12) and total costs are as follows: $G + R_I$ = €17,901.12, $G + R_{II}$ = €14,701 and $G + R_{III}$ = €15,894.13.

As we have just shown, the costs faced by an individual with moderate dependence may be higher than those that a dependent of severity level II or III has to afford, once he receives a public benefit. This is so because the amount of benefits received by individuals with dependence degree I is too small in comparison with the long-term care costs that this severity involves. Home care assistance provided by a professional has a cost which does not take advantage of the economy of scale compared to assistance in a medical centre or a residence.

Using data from the EDAD survey [17], and the methodology presented in Sect. 2, we have obtained the individual illiquidity risk by age and gender, and by the degree of severity (Table 5 and Fig. 2).

The results show three main facts: (i) illiquidity risk increases with age, (ii) in high ages the illiquidity risk in degree III increases faster than in other situations, (iii) women have higher illiquidity risk than men for any age and severity risk. These results are closely related to the evolution of the dependence rates. Women have higher illiquidity risk than men due both to physical and economic factors. Regarding physical factors, women have higher prevalence rates than men, and consequently the illiquidity risk is higher for them than for males. Moreover, since women live longer than men in any given dependence state, they may be more vulnerable. Regarding the economic factors, women receive lower pensions than men for many reasons that have been mentioned before, but essentially because they have contributed less than men to the social security pension system and so their pensions are lower. The fact that women live longer in a dependence state linked to low public pensions, increases the illiquidity risk for women compared to men, thus women have to face higher difficulties to meet long-term care costs. For instance, a men aged 79 has a 5 % risk of not having enough cash to afford care costs with moderate dependence, whereas for a women in dependence of degree I a risk of illiquidity equal to 5 % is already found at age 72, so seven years before men.

We have carried out a comparative analysis by establishing different scenarios according to the amount of pension that an individual may receive and his/her care costs depending on the degree of dependence. Table 6 shows the difference between an average pension income and expenditure for an individual by gender, by type of pension and by degree of severity of dependence. Females who receive an average retirement pension, have a clear disadvantage compared to males in the same situation, for any given degree of severity. This is due to the fact that women have average pensions lower than men (€9,654.12 per year versus €15,897.14). We have not considered differences in the care costs for men and women, respectively. Despite the type of pension scenario, we observe that for both genders we obtain a negative net result in cases of moderate dependence. This reveals that the most vulnerable financial situation

Table 5. Individual illiquidity risk by age, gender and dependence severity level.

Age	Males			Females		
	Degree I	Degree II	Degree III	Degree I	Degree II	Degree III
65	2.51 %	4.74 %	3.57 %	3.50 %	6.64 %	4.82 %
70	2.92 %	5.76 %	4.30 %	4.65 %	9.04 %	6.61 %
75	3.94 %	8.24 %	6.06 %	6.64 %	13.93 %	10.12 %
80	5.69 %	13.14 %	9.57 %	9.28 %	22.29 %	16.04 %
85	7.85 %	21.59 %	15.70 %	11.93 %	34.07 %	24.54 %
90	9.75 %	34.68 %	25.22 %	13.39 %	47.41 %	34.26 %
95	10.35 %	52.16 %	37.94 %	12.90 %	58.23 %	41.25 %

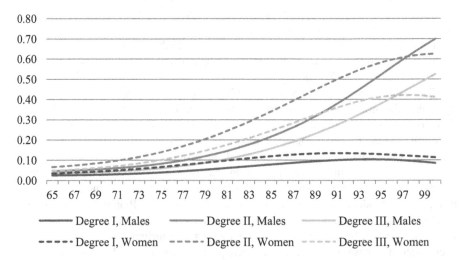

Fig. 2. Probability of illiquidity by age, gender and dependence severity level.

Table 6. Individual net amount by gender and dependence degree of severity by type of pension.

Pension	Males			Females		
	Degree I	Degree II	Degree III	Degree I	Degree II	Degree III
Retirement	−2,003.98	1,196.14	3.01	−8,247.00	−5,046.88	−6,240.01
Widowhood	−9,040.52	−5,840.40	−7,033.53	−9,040.52	−5,840.40	−7,033.53

Net amount expressed in euros.

that an individual could face is precisely at the moderate degree of dependence, because the benefit received from the public long-term care system is clearly insufficient (€3,600 yearly) in comparison to the costs that this state involves (€15,111 yearly). In the case of receiving only a widowhood pension both genders obtain negative results, but since men have a lower average pension of widowhood than women, their net situation is worse.

5   Conclusions

Moderate dependence is a situation with a low level of coverage by the long-term care public system in Spain. People who suffer this degree of dependence, in spite of the fact that they have more capacity to develop basic daily activities than dependent people with higher degree, they are more unprotected in financial terms. In the scenarios studied in this paper, dependence cost in degree I may be higher than long-term care costs related to more severe dependence, due to the fact that public benefits for this case are clearly insufficient. However, since prevalence rates of dependence increase with age in the higher degrees of severity, the illiquidity risk also increases with age in the elderly population as a whole.

Women are exposed to a higher illiquidity risk than men. They have higher life expectancy as well as higher prevalence rates in dependence, and receive lower retirement pensions than men. Consequently, women are more vulnerable than men to face unexpected costs such as long-term care costs and therefore they may be closer to a weak position from the financial point of view.

Acknowledgements. We thank the Spanish Ministry of Economy FEDER grants ECO2013-48326-C2-1-P and ECO2015-66314-R, and ICREA Academia.

References

1. OECD: Old-age income poverty. In: Pensions at a Glance 2011: Retirement-Income Systems in OECD and G20 Countries, OECD Publishing, Paris (2011). http://dx.doi.org/10.1787/pension_glance-2011-28-en
2. Bolancé, C., Alemany, R., Guillen, M.: Sistema público de dependencia y reducción del coste individual de cuidados a lo largo de la vida. Revista de Economía Aplicada **61**, 97–117 (2013)

3. Oliveira Martins, J., de la Maisonneuve, C.: The drivers of public expenditure on health and long-term care: an integrated approach. In: OECD Economic Studies, No. 43, February 2006
4. Guillen, M., Comas-Herrera, A.: How much risk is mitigated by LTC protection schemes? A methodological note and a case study of the public system in Spain. Geneva Pap. Risk Insur. - Issues Pract. **37**, 712–724 (2012)
5. Abellán García, A., Pujol Rodríguez, R.: Un perfil de las personas mayores en España, 2015. Indicadores estadísticos básicos. Informes envejecimiento en-red. Enero 2015, 10. Ministerio de Economía y Competitividad (2015)
6. Guillen, M., Albarrán, I., Alcañiz, M., Ayuso, M., Blay, D., Monteverde, M.: Longevidad y dependencia en España. Consecuencias sociales y económicas. Fundación BBVA, Madrid (2006)
7. Fernández, J.L., Forder, J., Truckeschitz, B., Rokosova, M., McDaid, D.: How Can European States Design Efficient, Equitable and Sustainable Funding Systems for Long-Term Care for Older People? Policy Brief 11. World Health Organisation Europe, Copenhagen (2009)
8. Chinchilla, N., Jiménez, E., Grau, M.: Impacto de las pensiones en la vejez. Jubilación y calidad de vida en España. VidaCaixa e IESE (2014)
9. Doménech, R.: Pensiones, bienestar y crecimiento económico. Informes Instituto BBVA de Pensiones, 14 March 2014
10. Comité de Expertos: Informe del Comité de Expertos sobre el factor de sostenibilidad del sistema público de pensiones (2013)
11. Hidalgo Vega, A., Calderón Milán, M.J., Pérez Camarero, S.: Composición y estructura de los hogares sustentados por personas mayores. Documento Universidad de Castilla-La Mancha (2008)
12. Inverco: Las Instituciones de Inversión Colectiva y los Fondos de Pensiones. Informe 2014 y perspectivas 2015 (2015)
13. Costa-Font, J., Gil-Trasfí, J., Mascarilla-Miró, O.: Capacidad de la Vivienda como Instrumento de Financiación de las Personas Mayores en España. Premio Edad & Vida 2005. Fundación Edad & Vida (2005)
14. Banco de España: Encuesta Financiera de las Familias (EFF) 2011: métodos, resultados y cambios desde 2008 (2011)
15. Jiménez, S., Nicodemo, C., Raya, J.M.: El diferente impacto del género en el Sistema de pensiones español. Ministerio de Empleo y Seguridad Social, Fondo de Investigación de la Protección Social (2010)
16. Alaminos, E., Ayuso, M.: Una estimación actuarial del coste individual de las pensiones de jubilación y viudedad: Concurrencia de pensiones del Sistema de la Seguridad Social español. Estudios de Economía Aplicada **33**(3), 817–838 (2015)
17. Instituto Nacional de Estadística (INE): Encuesta sobre Discapacidades, Autonomía personal y situaciones de Dependencia 2008 (EDAD 2008) (2009)
18. De Montesquieu, L.: Construction of Biometric Actuarial Bases for Long-Term Care Insurance. Technical report, SCOR Global Risk Center (2012)
19. Haberman, S., Pitacco, E.: Actuarial Models for Disability Insurance. Chapman & Hall/CRC, Boca Raton, USA (1999)
20. Pitacco, E.: Health Insurance. Basic Actuarial Models. EAA Series. Springer, Heidelberg (2014)
21. OECD: Pensions at a Glance 2013: OECD and G20 Indicators, OECD Publishing, Paris (2013). http://dx.doi.org/10.1787/pension_glance-2013-en

22. Ayuso, M., Guillen, M.: El coste de los cuidados de larga duración en España bajo criterios actuariales: ¿es sostenible su financiación? In: El Estado de bienestar en la encrucijada: nuevos retos ante la crisis, Ekonomi Gerizan, Federación de Cajas de Ahorro Vasco-Navarras, pp. 213–227 (2011)

23. Albarrán, I., Alonso, P., Bolancé, C.: Comparación de los baremos español, francés y alemán para medir la dependencia de las personas con discapacidad y sus prestaciones. Revista Española de Salud Pública **83**(3), 379–392 (2009)

24. Artís, M., Ayuso, M., Guillen, M., Monteverde, M.: Una estimación actuarial del coste individual de la dependencia en la población de mayor edad en España. Estadística Española **49**(165), 373–402 (2007)

Predicting Probability of Customer Churn in Insurance

Catalina Bolancé, Montserrat Guillen,
and Alemar E. Padilla-Barreto[✉]

Department of Econometrics, Riskcenter-IREA, University of Barcelona,
Av. Diagonal 690, 08034 Barcelona, Spain
{bolance,mguillen,alemarpadillabarreto}@ub.edu

Abstract. We focus on a real case of the motor insurance sector. We propose four different methods to predict lapsing from a portfolio of policies. We present a comparative analysis between three different performance measures in order to assess the predictive power of each model. Our comparison analyses the outcomes of a logistic regression, a conditional tree, a neural network and a support vector machine. These are all considered basic approaches to data mining and knowledge discovery. The main contribution of this paper is to show that, depending on the type of analysis and the objective of the researcher, the optimal prediction method may differ.

Keywords: Loyalty · Classification methods · ROC curve · Performance measures · Predictive analytics

1 Introduction

Customer loyalty is one of the most important priorities in most insurance companies. Different researchers argue that retaining a customer instead of getting a new one is a profitable strategy for the company, because the cost of finding a new customer is larger than the cost of keeping an existing one [1–3]. In fact, long-term customers tend to be more engaged with the company [4] and take less of the company's time than attracting new customers [5, 6].

Every day policyholders decide to leave the company and switch to a competitor. Customer satisfaction has a direct link with future revenue streams [7] and can be affected among other things by not investing in facilitating complaints of dis-satisfied customers [3], increased premiums, poor claims management, delays in payment of claims and firm performance. So, finding the reasons of customer churn and identifying factors that explain the switching behaviour [8] is a key aspect for the correct management and strategy of an insurance firm. The need for implementing short-term actions oriented to improve customer satisfaction and reverse the intention to leave is the reason for modelling and predicting the probability of churn through predictive analytical models [9–13].

The literature on customer churn does not necessarily compare the criteria used to select between modelling alternatives. In this paper, we propose performance measures and threshold values (cut-offs) that can guide practitioners in order to be able to decide

R. León et al. (Eds.): MS 2016, LNBIP 254, pp. 82–91, 2016.
DOI: 10.1007/978-3-319-40506-3_9

which is the best model in their particular context. We present an illustration from a Spanish motor insurance data.

In the case study that we present in the current paper, we are interested in the renewal probabilities of each customer in the insurance company for one sample. Therefore, we select the following classical predictive models (or classifiers): logistic regression, decision tree, neural networks and support vector machine. Those methods allow to classifying each new client using the probability of belonging to the group of customers that stay versus those who churn. In fact, we consider two differentiated groups in the classifying algorithm: retainers and churners.

We conduct our analysis by comparing the receiver operating characteristics (ROC) curves obtained in every method [14]. ROC curves are one of the most common tools for evaluating and comparing classifications models [15]. However, when ROC curves crossing each other once or more times, choosing the best method from this comparison may provide an ambiguous result [16]. Our principal contribution is devoted to looking for different adjustment criteria that compare the results for each of the above methods.

In this paper, we consider three kinds of criteria, two of them based on optimal cut-offs and another one based on the area under an ROC curve (AUC). We want to choose the best model according to the company's specific goals, based on these criteria. This fact, this has a direct implication on the predictive ability of the models, that is, different criteria provide different preference for models.

We use a large database of an insurance company, with information about the behaviour of the clients from the past. The methods and performance measure are implemented using version 3.2.3 of R language.

This paper is organized as follows: Sect. 2 gives a brief presentation of the different methods used to estimate the probability of belonging to each group and presents the discriminative measures based on thresholds and the ROC curve. Section 3 briefly describes the data and its pre-processing. In Sect. 4 we show the comparative analysis and discuss the results. And finally, in Sect. 5 we draw our conclusions.

2   Churn Prediction Models

Let $y_i = \{0, 1\}$ be an observation of the outcome random variable Y_i which provides information about the client decision. Whenever the customer stays in the company, i.e. renews the policy, the observed value is equal to 1 and when he churns, i.e. does not renew his policy, the observed value is equal to 0. Moreover, $X_i' = (x_{0i}, x_{1i}, \ldots, x_{ni})$ is a vector of n covariates, all of them associated with the ith policy. The total number of clients is denoted by N.

The churn prediction models are designed from a binary classification perspective, so, instances are assigned to one of the two classes Y = {Renew (or No Churn), Not renew (or Churn)} on the basis of their observed features X's. Classifiers discriminate instances from the former classes. In our case, we use soft classifiers [17, 18]. These methods estimate class probability p_i, which denote the probability that the ith policy churns, that allow to define the predicted classes by comparing p_i with different

Table 1. Confusion matrix for a given threshold t.

		Predicted	
		Not renew	Renew
Real	Not renew	True negative	False positive
	Renew	False negative	True positive

Source: Own design.

classification cut-offs $t \in [0, 1]$. Each of these comparisons produces a confusion matrix as shown in Table 1.

We present four popular techniques used for binary classification problems: logistic regression (LR), decision tree (DT), neural networks (NN) and support vector machine (SVM). These methods can generate a score [0, 100], so we can classify every customer based on this score. In fact, the score itself is just a transformation of the probability estimation that is obtained by each method.

2.1 Logistic Regression

Logistic regression model is a particular case of a generalized linear model. The specification of a generalized lineal model consists of three elements: a random variable Y, that takes value Y = 1 if customer renews the contract and takes value Y = 0 if customer does not renew the contract; a systematic linear predictor ($\eta = X'\beta$) component and a link function (g) that relates the expected value of the response variable with the linear predictor η. Under the logistic regression model we assume that the probability of renewal is:

$$P(Y = 1|X) = p = \frac{1}{1 + \exp(-X'\beta)},$$

where β is the parameter vector to be estimated.

Logistic regression is recognized for being simple to understand, it is appropriate when the dependent variable has only two possible values and it is easy to implement. If there are interactions between covariates, then interpretation is more complex and may lead to predictions that are too accurate (overfitting) in the sample, but may not be out of the sample. This is the case in general regression models, when too many parameters are entered into the model. Also, we have to consider that logistic regression is a parametric model that assumes a specific functional form in the link function and a particular distribution function in the exponential family.

2.2 Decision Trees

In our case, we design a conditional inference tree using the **ctree** function from the **party** package. This is a recursive partitioning algorithm that takes into account the concept of statistical significance, i.e., first, in each node a test is performed for the global hypothesis of independence between the variable response and any of the covariates, so, if the null hypothesis is rejected the significant association between Y and each inputs X's is evaluated and second a binary split is implemented once the association has been identified, using the variable with the strongest association to the dependent variable Y (see [19], for more details).

Decision trees are a powerful visuals tool that is able to properly handle the interaction between variables. The method is useful to form clusters, it is easy to understand, interpret and explain. Nevertheless if the training data set is somewhat different from the testing data, the results might not be robust. In fact, this turns out to be very similar to the overfitting phenomenon just described for the logistic regression. In addition, the **ctree** function allows working with variables of different nature. Therefore, transformations in intervals are not necessary. However, we need to consider that regression trees provide results that are difficult to interpret when the explanatory variables have different scales. Furthermore, within the **ctree** function partition criteria and algorithms can be based on different statistics such as the Gini coefficient or the deviance, among others.

2.3 Neural Networks

Neural networks are information processing systems inspired in biological neural associations. They are composed by interconnected neurons whose communication links are weighted. Generally, neural networks representations consist of input nodes, numeric weights, transfer functions, activation functions and output nodes, so they are able to represent non-linear mappings from input to output variables [20].

Neural networks can be trained to generate optimum results from the desired inputs, for example they can minimize false positives. However, neural networks are a sort of black box, making qualitative interpretation rather difficult. Moreover, overfitting and the adjustment of the parameters that control the algorithm is mandated by the criterion imposed by the network designer.

In this study we use the **nnet** function from the **nnet** package for adjust a neural network. This function fits a single-hidden-layer neural network based on a feed-forward neural network [21]. The overfitting is also a weakness of the neural networks approach.

2.4 Support Vector Machines

Support vector machine methods were first introduced by Boser et al. [22] and are useful for classification and regression analysis. These methods are based on quadratic optimization and allow to consider no linearity using different kernel functions to map data set samples into a higher dimensional space [23], looking for the hyperplane that optimally separates the data groups [24]. The main disadvantage is that support vector machines may also be difficult to interpret, although they are practical in terms of producing scores. Moreover, it has been demonstrated that the support vector machine approach is not accurate when the number of individual in each class is unbalanced (see, for example, [25]).

We applied the **svm** function from the **e1071** package [26] based on the LIBSVM library [27], considering one-classification type and a polynomial kernel.

2.5 Model Performance Measures

In order to evaluate the classifiers performance in defection prediction, we consider three different criteria based on: (C1) sensitivity plus specificity, (C2) accuracy for a specific threshold and (C3) the area under the curve (AUC). These measures summarize in a single scalar value the performance of each model. Besides, true and false positive cases are denoted as TP and FP, while true and false negatives are identified as TN and FN, respectively. Thus,

$$C1 = \max (\text{sensitivity} + \text{specificity}) = \max (TP/(TP + FN) + TN/(TN + FP)$$
$$C2 = \max (\text{accuracy}) = \max ((TP + TN)/(TP + TN + FP + FN))$$
$$C3 = \max (AUC)$$

The Area Under the Curve is a well-known measure to evaluate the discriminative power of the predictive models. An area under the diagonal in the ROC space equals 0.5 and it is associated to a pure random classification model, so the greater AUC, the better the model.

3   Data

The data are provided by a Spanish insurance company and included many variables that are expected to explain customer churn. This data base may include clients who have one or more motor insurance policies. It includes variables related to the policy, the customer and the vehicles as shown in Table 2. Generally, motor insurance policies are renewed every year. So, for our purposes, date of policy start, cancellation date and status of each policy are key drivers during the whole analysis.

The study is conducted in the usual two phases, one related to the pre-processing of data called filtering and another linked directly to the modelling process. Data pre-processing is a long but necessary process previous to the model implementation,

Table 2. Some variables used in the case study on modeling churn in motor insurance.

Related to	Variable
Policyholder	Sex, Age, Number of policies in force, Total number of policies in force that are owned by the same customer in other lines of insurance, Sum of premiums (both canceled and in force), Highest premium paid, Maximum duration of all policies owned by the same customer
Policy	Policy status (C = canceled, I = in force), Guarantees, New premium, Old premium, Change in premium (new vs previous), Change in premium (new vs the first), Discounts applied or Bonus-Malus level
Vehicle	Vehicle's type, Main driver's age, Second driver (Yes, No), Total cost of claims, Power (<=90 or >90)

Source: Own data set to study customer churn in motor insurance, 2015.

so, it is necessary to perform validation of non-duplication of policies, redefinition of the status of the policies, analysis of missing values, treatment of outliers, categorization and re-categorization of some variables and so on. The final dataset for this illustration has about 80 % proportion of policies in force versus 20 % cancelled. From this dataset and regarding to the analysis of missing values, we decide to use complete-case analysis (for further details, [28]), this gives rise to a total of 14, 000 policies approximately with the same proportion as before.

The former data base is divided into two groups (training set and test set) in a ratio of 70 %–30 %, respectively. With the modelling results, we generate for both the training and the test set a confusion matrix for each possible threshold from 0 to 1 and we calculate the following valuation measures: sensitivity, specificity, false positive rate, false negative rate, sensitivity, specificity and accuracy. Furthermore, we compute the AUC as before for the training and the test data set.

4   Results

We present the results based on the ROC curve and the different criteria. We show the criteria values and the related thresholds in the case of C1 and C2. Figure 1 shows each of the ROC curves associated to the classifiers for the test sample. We can observe multiple intersections, so, we do not observe a curve that dominates over the others. In fact, for each intersection we have different levels of false alarms where one classifier outperforms the others. Furthermore, the AUC index is located in the interval [0.88, 0.90], so a 0.20 probability of misclassification is obtained.

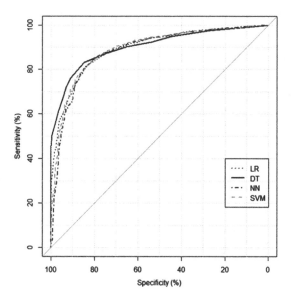

Fig. 1. Intersecting ROC curves

Table 3 shows the results using the training data set. In general, for each criterion we find small differences between the performance measure values. Neural networks model has the highest values in all criteria. Then, the biggest or lowest threshold is not necessarily in accordance with the best or worst value for both C1 and C2. Last, the largest area under the curve corresponds to the neural networks approach, followed by conditional tree, support vector machine and logistic regression.

Similarly, Table 4 presents the results for the test sample. We observe similar values for the performance measures compared to the results obtained in the training data set. In this case, the conditional tree has the highest performance measure for C1 and C3. In addition, in terms of C2 the support vector machine has a better performance. As before, a higher threshold does not necessarily mean a better performance measure or vice versa.

Table 3. Performance criteria for each model in the training data set.

| | | Model | | | |
		Logistic regression	Conditional tree	Neural network	Support vector machine
Criterion	C1	1.67	1.69	1.72	1.68
	C2 (%)	86.19	85.36	88.72	86.24
	C3 (%)	90.23	90.54	92.71	90.34
Optimal threshold	C1	0.75	[0.68,0.86]	0.77	0.82
	C2	0.55	[0.45,0.51]	[0.49,0.52]	0.57

Source: Own calculations

Table 4. Performance criteria for each model in the test data set.

| | | Model | | | |
		Logistic regression	Conditional tree	Neural network	Support vector machine
Criterion	C1	1.65	1.68	1.65	1.65
	C2 (%)	86.08	85.28	86.25	86.82
	C3 (%)	89.47	90.28	88.51	89.33
Optimal threshold	C1	0.77	[0.68,0.86]	0.82	0.84
	C2	0.48	[0.45,0.51]	0.50	0.52

Source: Own calculations

5 Conclusions

We have estimated churn probabilities using three different models, so we can predict the probability of renewal of a motor insurance policy. These probabilities give us a first scoring model in each case, so, it is possible to design marketing actions, aimed at specific customers. Selecting the best model, is not a trivial choice, but depends on the insurer preferences. We analyse the predictive performance of four modelling approaches. These are different alternatives that focus on similar, yet not exactly the same values. So, maximizing the number of true positive and negative cases, finding the greatest possible number of successes regarding non-churners and churners classification or just selecting the largest area under the curve, i.e. relying only, on the true positive and false positive rate, do not lead to the same conclusion on which is the best predictive approach.

In general, however, we obtain similar results for all the models considered here. Focusing on the results, we select neural networks as the best model for the training data and the conditional tree for the test data if we consider the first criterion (C1). Considering the second and third (C2 and C3) criteria, the neural network has the best performance for the training data. However, when looking at the test data, conditional trees perform better than the others for C3, while support vector machine performs better for C2.

Our results are useful for companies, since this illustration shows how the performance of models can be evaluated and how the modelling process can be assessed when analysing loyalty. In terms of limitations of the study we should mention that the predictive statistics described here could have studied together with a wider range of performance indicators. One of the practical implications of the paper is that no model seems to dominate the others, showing that there should not be a clear and general preference for any of the modelling alternatives presented here.

Further research should include additional performance diagnostics in order to evaluate the results of the predictive model. On the other hand, we suggest to improve the missing value treatment, for example, for unordered categorical predictors we recommend to add an additional category representing missing responses. Finally, we think that exploring the reactivity of the selected model taking into account profitability and the cost of each decision could be an interesting area of analysis.

Acknowledgements. We thank the Spanish Ministry of Economy FEDER grant ECO2013-48326-C2-1-P, AGAUR, and ICREA Academia.

References

1. Zeithaml, V.A., Berry, L.L., Parasuraman, A.: The behavioral consequences of service quality. J. Mark. **60**, 31–46 (1996)
2. Fornell, C., Wernerfelt, B.: Defensive marketing strategy by customer complaint management: a theoretical analysis. J. Mark. Res. **24**, 337–346 (1987)
3. Fornell, C., Wernerfelt, B.: A model for customer complaint management. Mark. Sci. **7**(3), 287–298 (1988)

4. Thuring, F., Nielsen, J.P., Guillen, M., Bolance, C.: Selecting prospects for cross-selling financial products using multivariate credibility. Expert Syst. Appl. **39**(10), 8809–8816 (1988)

5. Keaveney, S.M.: Customer switching behavior in service industries: an exploratory study. J. Mark. **59**, 71–82 (1995)

6. Brockett, P.L., Golden, L.L., Guillen, M., Nielsen, J.P., Parner, J., Pérez-Marín, A.M.: Survival analysis of a household portfolio of insurance policies: how much time do you have to stop total customer defection? J. Risk Insur. **75**(3), 713–737 (2008)

7. Fornell, C.: A national customer satisfaction barometer: the Swedish experience. J. Mark. **56**, 6–21 (1992)

8. Guillen, M., Nielsen, J.P., Scheike, T.H., Pérez-Marín, A.M.: Time-varying effects in the analysis of customer loyalty: a case study in insurance. Expert Syst. Appl. **39**(3), 3551–3558 (2012)

9. Guillen, M., Nielsen, J.P., Pérez-Marín, A.M.: The need to monitor customer loyalty and business risk in the European insurance industry. In: Geneva Papers on Risk and Insurance, Issues and Practice, pp. 207–218 (2008)

10. Guelman, L., Guillen, M.: A causal inference approach to measure price elasticity in automobile insurance. Expert Syst. Appl. **41**(2), 387–396 (2014)

11. Guelman, L., Guillen, M., Pérez-Marín, A.M.: A survey of personalized treatment models for pricing strategies in insurance. Insur. Math. Econ. **58**, 68–76 (2014)

12. Guelman, L., Guillen, M., Pérez-Marín, A.M.: A decision support framework to implement optimal personalized marketing interventions. Decis. Support Syst. **72**, 24–32 (2015)

13. Guelman, L., Guillen, M., Pérez-Marín, A.M.: Uplift random forests. Cybern. Syst. **46**(3–4), 230–248 (2015)

14. Guillen, M.: Regression with categorical dependent variables. In: Predictive Modeling Applications in Actuarial Science. Predictive Modeling Techniques, vol. 1, Cambridge University Press, Cambridge (2014)

15. Fawcett, T.: An introduction to ROC analysis. Pattern Recognit. Lett. **27**, 861–874 (2006)

16. Gigliarano, C., Figini, S., Muliere, P.: Making classifier performance comparisons when ROC curves intersect. Comput. Stat. Data Anal. **77**, 300–312 (2014)

17. Stripling, E., Vanden Broucke, S., Antonio, K., Baesens, B., Snoeck, M.: Profit maximizing logistic regression modeling for customer churn prediction. In: IEEE International Conference on Data Science and Advanced Analytics (DSAA), pp. 1–10 (2015)

18. Liu, Y., Zhang, H.H., Wu, Y.: Hard or soft classification? Large-margin unified machines. J. Am. Stat. Assoc. **106**(493), 166–177 (2011)

19. Hothorn, T., Hornik, K., Zeileis, A.: Unbiased recursive partitioning: a conditional inference framework. J. Comput. Graph. Stat. **15**(3), 651–674 (2006)

20. Bishop, C.M.: Neural Networks for Pattern Recognition. Oxford University Press, Oxford (1995)

21. Ripley, B.D.: Pattern Recognition and Neural Networks. Cambridge University Press, Cambridge (1996)

22. Boser, B.E., Guyon, I.M., Vapnik, V.N.: A training algorithm for optimal margin classifiers. In: Proceedings of the Fifth Annual Workshop on Computational Learning Theory, pp. 144–152. ACM (1992)

23. Karatzoglou, A., Meyer, D., Hornik, K.: Support vector machines in R. J. Stat. Softw. **15**(9), 1–28 (2006)

24. Hsu, C.W., Chang, C.C., Lin, C.J.: A Practical Guide to Support Vector Classification. Department of Computer Science, National Taiwan University, Taiwan (2003)

25. Tao, D., Tang, X., Li, X., Wu, X.: Asymmetric bagging and random subspace for support vector machines-based relevance feedback in image retrieval. IEEE Trans. Pattern Anal. Mach. Intell. **28**(7), 1088–1099 (2006)
26. Meyer, D., Dimitriadou, E., Hornik, K., Weingessel, A., Leisch, F.: Package "e1071". Misc. Functions of the Department of Statistics (e1071), TU Wien. The comprehensive R archive network (2012)
27. Chang, C.C., Lin, C.J.: LIBSVM: a library for support vector machines. ACM Trans. Intell. Syst. Technol. (TIST) **2**(3), 27 (2011)
28. Gelman, A., Hill, J.: Missing-data imputation. Data Analysis Using Regression and Multilevel/Hierarchical Models, 1st edn, pp. 529–544. Cambridge University Press, Cambridge (2006)

Integrated Supply Chain Planning: A Review

Herminia I. Calvete[1], Carmen Galé[2(✉)], and Lorena Polo[3]

[1] Dpto. de Métodos Estadísticos, IUMA, Universidad de Zaragoza,
Pedro Cerbuna 12, 50009 Saragossa, Spain
herminia@unizar.es
[2] Dpto. de Métodos Estadísticos, IUMA, Universidad de Zaragoza,
María de Luna 3, 50018 Saragossa, Spain
cgale@unizar.es
[3] División Logística, Instituto Tecnológico de Aragón,
María de Luna 7-8, 50018 Saragossa, Spain
lpolo@itainnova.es

Abstract. Supply chain management requires making strategic, tactical and operational decisions which involve suppliers, manufacturers and customers. Potential decision makers are involved in a centralized/decentralized decision structure which describes their relationships and interdependencies. The use of mathematical models to optimize supply chain planning problems one at a time has been extensively discussed in the literature. The integration of two or more of these problems and the use of mathematical programming to model them is less frequent. This paper reviews the main mathematical models recently developed to address the decision making process in integrated deterministic supply chain planning, emphasizing the processes which are integrated and the characteristics of the model.

Keywords: Supply chain management · Integrated decision-making process · Mathematical program · Optimization

1 Introduction

Nowadays, the problems involved in the efficient management of a supply chain (SC) constitute a significant area of research. A glance at the wide range of publications on supply chain management (SCM) reveals that this subject has been widely studied from different perspectives and disciplines in the literature. Schneeweiss [1] provides a review of how decision-making problems in SCM are differently managed from the perspectives of applied mathematics, operations research, economics and artificial intelligence. From whichever point of view, the current cornerstone of SCM is to consider the supply chain globally rather than as a set of individual activities. This implies identifying the logistic activities, recognizing the organizations and/or decision makers involved in the SC and integrating them in an efficient manner. The challenge in SC integration is to coordinate the activities involved so that the performance can be improved by reducing costs, increasing service levels, providing a better use of resources and offering an effective response to the changes in the market place [2].

© Springer International Publishing Switzerland 2016
R. León et al. (Eds.): MS 2016, LNBIP 254, pp. 92–103, 2016.
DOI: 10.1007/978-3-319-40506-3_10

The aim of this paper is to identify the mathematical programming approaches proposed in the literature to model the relationship amongst SC agents and their interdependencies. We pay attention to papers which use deterministic mathematical programming models when dealing with an integrated problem in SCM. We have searched under "supply chain management" in the available databases Scopus, Sciencedirect, ProQuest and Google Scholar. The results show an increasing number of works on SCM in recent years. Based on these results we have selected papers which include either in the abstract or the keywords the terms program, mathematical or algorithm. The paper is organized as follows. Section 2 explores the literature on SCM and identifies the classification criteria proposed in previous reviews of integrated SC problems. Sections 3 to 6 go on to discuss recent reviews on the subjects as well as interesting papers not included in these reviews. Finally, Sect. 7 sets out our findings and conclusions as well as areas of interest for further research.

2 Classification Criteria

When analyzing SCM, Beamon [3] distinguishes between the production planning and inventory control process and the distribution and logistics process. Both kinds of process interact with each other to produce an integrated SC. Min and Zhou [4] state that a SC consists of two main business processes: material management (inbound logistics) and physical distribution (outbound logistics). These processes consist of three levels of decision hierarchy: competitive strategy, tactical plans and operational routines. Location-allocation decisions and demand planning are included in the first level, which can be considered as long-term planning. Tactical problems or medium-term planning are the inventory control and the production/distribution coordination. Finally, vehicle routing/scheduling is considered as part of the operational routines or short-term planning. Goetschalckx et al. [5] review mathematical programming models for the efficient design of global logistics systems that take into account strategic and tactical considerations. Melo et al. [6] review facility location models which also involve decisions about the design of the supply chain network. They include papers dealing with tactical and operational decisions such as capacity allocation, inventory, procurement, production, routing or transportation modes. Schmid et al. [7] focus on the classical vehicle routing problem but their search is extended to incorporate tactical and operational problems related to production, warehousing and inventory. Okongwu et al. [8] study the impact on the performance of the SC of the way in which firms plan tactical aspects such as procurement, production and distribution and argue in favor of their integration.

As a conclusion, the following logistic activities can be singled out in SCM: procurement, facility location, production, inventory management and transportation. Based on these and on the optimization models which have been proposed to deal with integrated SCM, we have classified the selected papers into four categories: location/routing, sourcing/production and supplier selection/inventory, location/ inventory and inventory/routing, and production/distribution. In what follows, we include for each category a description of the problem and explore the relevant literature. The recent reviews discussed provide an important set of references. Only those papers either missing from these reviews or published more recently are explicitly mentioned here.

3   Location/Routing Problem (LRP)

This problem is defined for a network with, in general, a homogeneous set of vehicles with limited capacities which are shared by all depots. The goal is to determine which depots must be opened and decide the routes for each open depot and its customers. Early papers dealing with this problem considered uncapacitated depots. However, in more recent papers capacity constraints on depots and vehicles are addressed. The comprehensive review by Nagy and Salhi [9] published in 2007 proposes a classification scheme that takes into account four key aspects: the structure of the problem which involves entities and their organization; the type of input data which can be deterministic or stochastic; the planning period which classifies papers on location/routing in single or multi-periods; and the solution method that distinguishes between exact or heuristic procedures. Prodhon and Prins [10] provide a literature review on the subject until 2013 in which extensions of the classical location/routing problems are included, such as multi-echelon problems, applications with multiple objectives or uncertain data. Drexl and Schneider [11] review papers until 2014 and consider only problems where the selection of the facilities is not implicitly determined by the routing decisions. Both recent reviews provide an updated bibliography on the subject.

Amongst the papers not included in the previous reviews, Amiri [12] proposes a mixed integer program which is solved using Lagrangean relaxation combined with a heuristic procedure. Ross and Jaramayan [13] formulate a strategic model as a binary integer program to determine the location and routing solution. Then, an executional model, which is a linear program, provides the amount of product shipped between nodes. To solve the model they integrate simulated annealing and tabu search. Olivares-Benitez et al. [14] generate approximate efficient sets for the biobjective mixed integer program which minimizes the cost of the location/routing problem and the longest transportation time from the plants to the customers. Borges Lopes et al. [15] propose an evolutionary algorithm whose chromosome represents a complete solution i.e. the collection of routes and uses local search in the mutation phase.

In general, the models proposed in the literature are binary integer programs or mixed integer programs with a single objective function which is the sum of the opening costs and the routing costs. A few papers in this category consider multiobjective optimization models. The objectives take into account the location or the operational costs as well as the covered demand, the work-load balance or the longest transportation time. In problems which involve hazardous waste management, a social cost is introduced which takes into account social rejection or transportation risk. A few papers propose exact approaches for solving the problem mainly based on branch-and-cut algorithms. For large-size instances, metaheuristics have been proposed aiming to find good quality solutions in short computing times. Thus, simulated annealing, tabu search, evolutionary algorithms or adaptive large neighborhood search metaheuristics have been developed.

4 Sourcing/Production Problem (SPP) and Supplier Selection/Inventory Problem (SSIP)

Purchasing is one of the most strategically activities in SCM because it provides the opportunity to reduce costs across the entire SC. Since the cost of raw materials represents a substantial percentage of the total product cost, an essential task in the purchasing process is supplier selection. In addition, a relevant problem in SCM is to determine the appropriate level of inventory at each stage involved in a SC. Both supplier selection and inventory management are closely related to the production problem. The production of many different products with short life cycles and the need to provide customers a better service at a reduced cost lead to the integration of production and inventory related decisions. Usually, the manufacturer purchases the raw material and/or semi-finished items from several preferred suppliers. These raw parts are stored at the manufacturing facility or transformed into final products, as they are requested to satisfy demand from retailers.

Aissaoui et al. [16] present a literature review until 2007 on supplier selection. They focus especially on the problem of determining the best mixture of suppliers and allocating orders so as to satisfy different purchasing requirements. Ben-Daya et al. [17] propose a nonlinear mixed integer model for the sourcing/production problem which is approximately solved by applying derivative-free methods. The single objective function is computed as the sum of the costs incurred by the supplier and the manufacturer (production setup cost, raw material ordering and holding costs, and finished product holding cost) and the retailers (ordering cost and finished product inventory holding cost). Sawik [18] proposes a mixed integer programming problem for determining the latest delivery dates of product-specific parts. Two different approaches are compared. In the monolithic approach, the manufacturing, supply and assembly schedules are determined simultaneously. In the hierarchical approach, first an assignment of customer orders over the planning horizon and thereby the finished products assembly schedule is determined, and then the parts manufacturing and the supply schedules are decided. Concerning the supplier selection/inventory problem, in general only the minimization of the total system costs is considered when tackling both the supplier selection and the lot size decision. Keskin et al. [19] and Ventura et al. [20] formulate mixed integer nonlinear programming models. Cárdenas-Barrón et al. [21] propose a mixed integer linear model which is solved with a reduce and optimize approach. Glock [22] presents a continuous mathematical model and suggests heuristic solution procedures. Ustun and Demirtas [23] propose a multiobjective mixed integer linear programming model whose aim is to maximize the total value of purchasing, to minimize the total cost and the total defect rate and to balance the total cost between periods. The most preferred efficient solutions are determined by considering the decision maker's preferences. Rezaei and Davoodi [24] propose two multiobjective mixed integer nonlinear models. The three objective functions are based on cost, quality and service level. They develop a multiobjective evolutionary algorithm to approximate the set of Pareto solutions. Huang et al. [25] model the problem as a three-level dynamic non-cooperative game. The authors propose both analytical and computational methods to compute the Nash equilibrium.

5   Location/Inventory Problem (LIP) and Inventory/Routing Problem (IRP)

Location/inventory problems involve integrated decisions on the location of facilities such as distribution centers or factories, and inventory management. The aim is to minimize transportation and facility fixed costs as well as inventory holding and handling costs. These problems are closely related to the inventory/routing problems which have received more attention in the literature. The inventory/routing problem can be seen as an extension of the vehicle routing problem where the aim is to satisfy the demand of customers while minimizing the total distribution cost. In the inventory/routing problem, the amount to supply to each customer (inventory allocation decision) as well as the time for delivery is determined in order to minimize the inventory cost. Moin and Salhi [26] present an overview of the area of inventory routing and classify papers according to the planning horizon: single-period, multi-period and infinite horizon models. Two more papers on this topic are worth mentioning. Tancrez et al. [27] propose a nonlinear continuous mathematical model with a single objective function obtained as the sum of periodic transportation, inventory holding, fixed operational costs and handling costs. Melo et al. [28] propose a mixed integer linear program whose single objective function accounts for the total net SC cost defined as the sum of supply, transportation and inventory holding costs and fixed costs for operating the facilities. In both papers a heuristic procedure is used to solve the problem.

The review by Andersson et al. [29] covers inventory management and routing from an operations research perspective until 2009. It takes into account the relation between science and practice by considering papers which consist of both case studies based on real applications and theoretical contributions based on idealized models. Coelho et al. [30] give a comprehensive review of inventory routing problems until 2013, by categorizing them with respect to their structural variants and the information available on customer demand. Yu et al. [31], Day et al. [32] and Coelho and Laporte [33] deal with mixed integer linear models. The papers by Berman and Wang [34] and Kuhn and Liske [35] propose nonlinear models. Song and Savelsbergh [36] give a different perspective and focus on how to measure the performance of the distribution strategies and also on which factors contribute to the complexity of the problem.

Most papers on this topic propose a mixed integer optimization model with a single objective function which reflects the overall costs. Branch-and-cut techniques are used to develop algorithms to exactly solve small-scale problems. When considering real-life instances, most algorithms are based on heuristics which take advantage of decomposing the problem into several sub-problems.

In this category we also include papers which focus on the strategic design of a SC network and deal with location, routing, production and inventory. The papers by

Ambrosino and Scutellà [37], Melo et al. [38], Hinojosa et al. [39], Thanh et al. [40], Manzini and Bindi [41], Sadjady and Davoudpour [42], Guerrero et al. [43], formulate mixed integer models with a single objective function aiming to operate the SC network at minimum cost.

6 Production/Distribution Problem (PDP)

Production and distribution planning are tactical decisions. In these problems, the assignment of available capacity in production sites, the inventory allocation to maintain a level of service of a retail or distribution facility and the distribution of the stocks to retailers have to be determined. Mula et al. [44] review mathematical programming models for SC production and transport planning until 2009. The authors propose eight classification criteria: supply chain structure, decision level, modeling approach, shared information, purpose, limitations, novelty and practical application. A later review until 2014 by Díaz-Madroñero et al. [45] propose a classification framework based on production, inventory and routing aspects, objective function structure and solution approach. Fahimnia et al. [46] provide a comprehensive review until 2010 and a classification of production/distribution models based on their degree of complexity and the level of simplification of the real-life scenario used. Adulyasak et al. [47] review until 2014 problem formulations and solution techniques to solve a production routing problem jointly optimizing production, inventory, distribution and routing decisions.

Other papers in this category include Elhedhli and Goffin [48], Manzini [49], Raa et al. [50] and Low et al., [51] which consider a single objective function accounting for the total cost. Liu and Papageorgiou [52] develop a multiobjective mixed integer linear model to simultaneously minimize total cost, total flow time and total lost sales. The authors apply the epsilon-constraint method and the lexicographic minimax method to find an efficient solution. Chan et al. [53] also consider several criteria (operating cost, order fulfillment lead time, and equity of utilization ratios) which are weighted to obtain a single objective function by using the analytic hierarchy process. Dawande et al. [54] present a study of conflict and cooperation aspects arising in SC when there is a conflict between the goals of the manufacturer and distributor. They conclude that the cost saving resulting from cooperation is usually significant. When collaborative planning is possible, Selim et al. [55] develop a multiobjective linear programming model which is solved by using a fuzzy goal programming approach. However, this collaboration is not always possible. Calvete et al. [56] focus on a hierarchical production/distribution planning problem that takes into account the existence of two decision makers who control the production and the distribution processes, respectively, and do not cooperate because of different optimization strategies. A bilevel mixed integer programming model is formulated for solving the problem. Amorin et al. [57] propose a multiobjective mixed-integer model to solve the integrated production and distribution planning of perishable goods which takes into account the value of freshness of the goods in addition to costs.

Most papers discuss either linear programming or mixed integer programming models. In contrast, nonlinear programming and multiobjective programming modeling approaches are scarcely applied in practice. Concerning the algorithmic approaches, only a few papers propose exact approaches, these being mainly based on branching processes improved with the introduction of valid inequalities or Lagrangean relaxation. Metaheuristic methods have also been proposed such as genetic algorithms, tabu search or ant colony optimization.

7 Findings, Conclusions and Further Research

In this paper we have reviewed the existing literature on integrated problems in SCM which connect several SC partners and which are related to strategic, tactical and operational decision making. The aim is to provide the researcher with a summary of the mathematical approaches which have been proposed to deal with these problems. The focus is on reviewing the manner in which the mathematical approach models the interdependencies among supply chain parties. Table 1 summarizes the findings. The analysis reveals that the mathematical models formulated in the papers are binary/integer/mixed-integer optimization problems, either linear or nonlinear. Most of them involve a single objective function which reflects the overall cost of the system and implicitly assumes a centralized decision making process. This mathematical approach allows the single decision maker to simultaneously solve all the problems involved. Only a few papers formulate multiobjective programs which allow us to take into account several conflicting objectives or the existence of several decision makers who are ready to collaborate. Even fewer papers introduce bilevel programming or game theory concepts to model the decision making process in decentralized systems without collaboration among partners. However, collaboration in the SC is not an easy task. Kampstra et al. [58] analyze the reality of the SC and describe the gap between the interest in supply chain collaboration and the relatively few recorded cases of successful application. Huang et al. [59] analyze the impact of sharing information on SC dynamics and demonstrate that understanding the advantages and limitations of each modeling approach can help researchers to choose the correct approach for studying their problem. It is worth pointing out that even in monolithic models, different mathematical programming formulations can capture the different points of view of agents in the SC such as customers and suppliers. Kalcsics et al. [60] illustrate how the mathematical program proposed depends on the decision maker the problem focuses on. Bertazzi et al. [61] also emphasize the impact of the objective function on the solution. It can reflect different decision policies and/or different decision makers, i.e. diverse organizational aspects of a company with a different emphasis on what is important in making a decision. In a decentralized decision making environment in which the SC agents operate independently in an organizational hierarchy, non-monolithic models such as bilevel programs are more appropriate. Cao and Chen [62] and Calvete et al. [56, 63] provide examples which show the differences between monolithic and non-monolithic models.

In our opinion, the mathematical approaches described in the literature under review (single objective programs, multiobjective programs and bilevel programs) contain the foundations of the deterministic optimization models that allow the integration of several processes of the SC involving either collaborative or non-collaborative decision

Table 1. Supply chain management problems and references in the paper.

Authors	References	Type of problem
Adulyasak, Y. et al. (2015)	[47]	PDP
Aissaoui, N. et al. (2007)	[16]	SPP and SSIP
Ambrosino, D. and Scutellà, M. (2005)	[37]	LIP and IRP
Amiri, A. (2006)	[12]	LRP
Amorim, P. et al. (2012)	[57]	PDP
Andersson, H. et al. (2010)	[29]	LIP and IRP
Ben-Daya, M. et al. (2013)	[17]	SPP and SSIP
Berman, O. and Wang, Q. (2006)	[34]	LIP and IRP
Borges Lopes, R. et al. (2016)	[15]	LRP
Calvete, H.I. et al. (2014)	[56]	PDP
Cárdenas-Barrón, L.E. et al. (2015)	[21]	SPP and SSIP
Chan, F. et al. (2005)	[53]	PDP
Coelho, L.C. et al. (2014)	[30]	LIP and IRP
Coelho, L. and Laporte, G. (2014)	[33]	LIP and IRP
Dawande, M. et al. (2006)	[54]	PDP
Day, J. et al. (2009)	[32]	LIP and IRP
Díaz-Madroñero, M. et al. (2015)	[45]	PDP
Drexl, M. and Schneider, M. (2015)	[11]	LRP
Elhedhli, S. and Goffin, J. (2005)	[48]	PDP
Fahimnia, B. et al. (2013)	[46]	PDP
Glock, C. (2011)	[22]	SPP and SSIP
Guerrero, W. et al. (2013)	[43]	LIP and IRP
Hinojosa, Y. et al.	[39]	LIP and IRP
Huang, Y. et al. (2011)	[25]	SPP and SSIP
Keskin, B. et al. (2010)	[19]	SPP and SSIP
Kuhn, H. and Liske, T. (2011)	[35]	LIP and IRP
Liu, S. and Papageorgiou, L. (2013)	[52]	PDP
Low, C. et al. (2014)	[51]	PDP
Manzini, R. (2012)	[49]	PDP
Manzini, R. and Bindi, F. (2009)	[41]	LIP and IRP
Melo, M. et al. (2005)	[38]	LIP and IRP
Melo, M. et al. (2012)	[28]	LIP and IRP
Moin, N. and Salhi, S. (2007)	[26]	LIP and IRP
Mula, J. et al. (2010)	[44]	PDP
Nagy, G. and Salhi, S. (2007)	[9]	LRP
Olivares-Benitez, E. et al. (2013)	[14]	LRP
Prodhon, C. and Prins, C. (2014)	[10]	LRP
Raa, B. et al. (2013)	[50]	PDP
Rezaei, J. and Davoodi, M. (2011)	[24]	SPP and SSIP
Ross, A. and Jayaraman, V. (2008)	[13]	LRP

(Continued)

Table 1. (*Continued*)

Authors	References	Type of problem
Sadjady, H. and Davoudpour, H. (2012)	[42]	LIP and IRP
Sawik, T. (2009)	[18]	SPP and SSIP
Selim, H. et al. (2008)	[55]	PDP
Song, J. and Savelsbergh, M. (2007)	[36]	LIP and IRP
Tancrez, J. et al. (2012)	[27]	LIP and IRP
Thanh, P. et al. (2008)	[40]	LIP and IRP
Ustun, O. and Demirtas, E. (2008)	[23]	SPP and SSIP
Ventura, J. et al. (2013)	[20]	SPP and SSIP
Yu, Y. et al. (2008)	[31]	LIP and IRP

makers. Nevertheless, it would be necessary to extend their use, especially the more complex models, in the context of real-system applications. It is worth mentioning the existence of a substantial number of papers devoted to SC configurations based on the concept of interdependence among partners in the SC. This contrasts with the scarce number of papers which use multiobjective optimization or hierarchical optimization for modeling the complexity of the decision processes involved in SCM. A future line of research should fill this gap and analyze the impact of using different mathematical approaches on the modeling of the SC performance.

Acknowledgments. This research work has been partially funded by the Gobierno de Aragón under grant E58 (FSE).

References

1. Schneeweiss, C.: Distributed decision making in supply chain management. Int. J. Prod. Econ. **84**, 71–83 (2003)
2. Simchi-Levi, D., Kaminsky, P., Simchi-Levi, E.: Designing and Managing the Supply Chain: Concepts, Strategies and Case Studies, 3rd edn. McGraw-Hill, New York (2009)
3. Beamon, B.: Supply chain design and analysis: models and methods. Int. J. Prod. Econ. **55**, 281–294 (1998)
4. Min, H., Zhou, G.: Supply chain modeling: past, present and future. Comp. Ind. Eng. **43**, 231–249 (2002)
5. Goetschalckx, M., Vidal, C., Dogan, K.: Modeling and design of global logistics systems: a review of integrated strategic and tactical models and design algorithms. Eur. J. Oper. Res. **143**, 1–18 (2002)
6. Melo, M., Nickel, S., da Gama, F.S.: Facility location and supply chain management - a review. Eur. J. Oper. Res. **196**, 401–412 (2009)
7. Schmid, V., Doerner, K., Laporte, G.: Rich routing problems arising in supply chain management. Eur. J. Oper. Res. **224**, 435–448 (2013)
8. Okongwu, U., Lauras, M., François, J., Deschamps, J.C.: Impact of the integration of tactical supply chain planning determinants on performance. J. Manuf. Sys. **38**, 181–194 (2016)

9. Nagy, G., Salhi, S.: Location-routing: issues, models and methods. Eur. J. Oper. Res. **177**, 649–672 (2007)
10. Prodhon, C., Prins, C.: A survey of recent research on location-routing problems. Eur. J. Oper. Res. **238**, 1–17 (2014)
11. Drexl, M., Schneider, M.: A survey of variants and extensions of the location-routing problem. Eur. J. Oper. Res. **241**, 283–308 (2015)
12. Amiri, A.: Designing a distribution network in a supply chain system: formulation and efficient solution procedure. Eur. J. Oper. Res. **171**, 567–576 (2006)
13. Ross, A., Jayaraman, V.: An evaluation of new heuristics for the location of cross-docks distribution centers in supply chain network design. Comp. Ind. Eng. **55**, 64–79 (2008)
14. Olivares-Benitez, E., Ríos-Mercado, R., González-Velarde, J.: A metaheuristic algorithm to solve the selection of transportation channels in supply chain design. Int. J. Prod. Econ. **145**, 161–172 (2013)
15. Borges Lopes, R., Ferreira, C., Sousa Santos, B.: A simple and effective evolutionary algorithm for the capacitated location-routing problem. Comp. Oper. Res. **70**, 155–162 (2016)
16. Aissaoui, N., Haouari, M., Hassini, E.: Supplier selection and order lot sizing modeling: a review. Comp. Oper. Res. **34**, 3516–3540 (2007)
17. Ben-Daya, M., Asad, R., Seliaman, M.: An integrated production inventory model with raw material replenishment considerations in a three layer supply chain. Int. J. Prod. Econ. **143**, 53–61 (2013)
18. Sawik, T.: Coordinated supply chain scheduling. Int. J. Prod. Econ. **120**, 437–451 (2009)
19. Keskin, B., Üster, H., Çetinkaya, S.: Integration of strategic and tactical decisions for vendor selection under capacity constraints. Comp. Oper. Res. **37**, 2182–2191 (2010)
20. Ventura, J., Valdebenito, V., Golany, B.: A dynamic inventory model with supplier selection in a serial supply chain structure. Eur. J. Oper. Res. **230**, 258–271 (2013)
21. Cárdenas-Barrón, L.E., González-Velarde, J.L., Treviño-Garza, G.: A new approach to solve the multi-product multi-period inventory lot sizing with supplier selection problem. Comp. Oper. Res. **64**, 225–232 (2015)
22. Glock, C.: A multiple-vendor single-buyer integrated inventory model with a variable number of vendors. Comp. Ind. Eng. **60**, 173–182 (2011)
23. Ustun, O., Demirtas, E.: An integrated multi-objective decision-making process for multi-period lot-sizing with supplier selection. Omega **36**, 509–521 (2008)
24. Rezaei, J., Davoodi, M.: Multi-objective models for lot-sizing with supplier selection. Int. J. Prod. Econ. **130**, 77–86 (2011)
25. Huang, Y., Huang, G., Newman, S.: Coordinating pricing and inventory decisions in a multi-level supply chain: a game-theoretic approach. Trans. Res. E **47**, 115–129 (2011)
26. Moin, N., Salhi, S.: Inventory routing problems: a logistical overview. J. Oper. Res. Soc. **58**, 1185–1194 (2007)
27. Tancrez, J., Lange, C., Semal, P.: A location-inventory model for large three-level supply chains. Trans. Res. E **48**, 485–502 (2012)
28. Melo, M., Nickel, S., da Gama, F.S.: A tabu search heuristic for redesigning a multi-echelon supply chain network over a planning horizon. Int. J. Prod. Econ. **136**, 218–230 (2012)
29. Andersson, H., Hoff, A., Christiansen, M., Hasle, G., Løkketangen, A.: Industrial aspects and literature survey: combined inventory management and routing. Comp. Oper. Res. **37**, 1515–1536 (2010)
30. Coelho, L.C., Cordeau, J.F., Laporte, G.: Thirty years of inventory-routing. Trans. Sci. **48**, 1–19 (2014)
31. Yu, Y., Chen, H., Chu, F.: A new model and hybrid approach for large scale inventory routing problems. Eur. J. Oper. Res. **189**, 1022–1040 (2008)

32. Day, J., Wright, P., Schoenherr, T., Venkataramanan, M., Gaudette, K.: Improving routing and scheduling decisions at a distributor of industrial gasses. Omega **37**, 227–237 (2009)
33. Coelho, L., Laporte, G.: Improved solutions for inventory-routing problems through valid inequalities and input ordering. Int. J. Prod. Econ. **155**, 391–397 (2014)
34. Berman, O., Wang, Q.: Inbound logistic planning: minimizing transportation and inventory cost. Trans. Sci. **40**, 287–299 (2006)
35. Kuhn, H., Liske, T.: Simultaneous supply and production planning. Int. J. Prod. Res. **49**, 3795–3813 (2011)
36. Song, J., Savelsbergh, M.: Performance measurement for inventory routing. Trans. Sci. **41**, 44–54 (2007)
37. Ambrosino, D., Scutellà, M.: Distribution network design: new problems and related models. Eur. J. Oper. Res. **165**, 610–624 (2005)
38. Melo, M., Nickel, S., da Gama, F.S.: Dynamic multi-commodity capacitated facility location: a mathematical modeling framework for strategic supply chain planning. Comp. Oper. Res. **33**, 181–208 (2005)
39. Hinojosa, Y., Kalcsics, J., Nickel, S., Puerto, J., Velten, S.: Dynamic supply chain design with inventory. Comp. Oper. Res. **35**, 373–391 (2008)
40. Thanh, P., Bostel, N., Péton, O.: A dynamic model for facility location in the design of complex supply chains. Int. J. Prod. Econ. **113**, 678–693 (2008)
41. Manzini, R., Bindi, F.: Strategic design and operational management optimization of a multi stage physical distribution system. Trans. Res. E **45**, 915–936 (2009)
42. Sadjady, H., Davoudpour, H.: Two-echelon, multi-commodity supply chain network design with mode selection, lead-times and inventory costs. Comp. Oper. Res. **39**, 1345–1354 (2012)
43. Guerrero, W., Prodhon, C., Velasco, N., Amaya, C.: Hybrid heuristic for the inventory location-routing problem with deterministic demand. Int. J. Prod. Eco. **146**, 359–370 (2013)
44. Mula, J., Peidro, D., Díaz-Madroñero, M., Vicens, E.: Mathematical programming models for supply chain production and transport planning. Eur. J. Oper. Res. **204**, 377–390 (2010)
45. Díaz-Madroñero, M., Peidro, D., Mula, J.: A review of tactical optimization models for integrated production and transport routing planning decisions. Comp. Ind. Eng. **88**, 518–535 (2015)
46. Fahimnia, B., Farahani, R., Marian, R., Luong, L.: A review and critique on integrated production-distribution planning models and techniques. J. Manuf. Sys. **32**, 1–19 (2013)
47. Adulyasak, Y., Cordeau, J., Jans, R.: The production routing problem: a review of formulations and solution algorithms. Comp. Oper. Res. **55**, 141–152 (2015)
48. Elhedhli, S., Goffin, J.: Efficient production-distribution system design. Manag. Sci. **51**, 1151–1164 (2005)
49. Manzini, R.: A top-down approach and a decision support system for the design and management of logistic networks. Trans. Res. E **48**, 1185–1204 (2012)
50. Raa, B., Dullaert, W., Aghezzaf, E.: A matheuristic for aggregate production-distribution planning with mould sharing. Int. J. Prod. Econ. **145**, 29–37 (2013)
51. Low, C., Chang, C.M., Li, R.K., Huang, C.L.: Coordination of production scheduling and delivery problems with heterogeneous fleet. Int. J. Prod. Econ. **153**, 139–148 (2014)
52. Liu, S., Papageorgiou, L.: Multiobjective optimisation of production, distribution and capacity planning of global supply chains in the process industry. Omega **41**, 369–382 (2013)
53. Chan, F., Chung, S., Wadhwa, S.: A hybrid genetic algorithm for production and distribution. Omega **33**, 345–355 (2005)
54. Dawande, M., Geismar, H.N., Hall, N., Chelliah, S.: Supply chain scheduling: distribution systems. Prod. Oper. Manag. **15**, 243–261 (2006)

55. Selim, H., Araz, C., Ozkarahan, I.: Collaborative production-distribution planning in supply chain: a fuzzy goal programming approach. Trans. Res. E **44**, 396–419 (2008)
56. Calvete, H.I., Galé, C., Iranzo, J.A.: Planning of a decentralized distribution network using bilevel optimization. Omega **49**, 30–41 (2014)
57. Amorim, P., Günther, H.O., Almada-Lobo, B.: Multi-objective integrated production and distribution planning of perishable products. Int. J. Prod. Econ. **138**, 89–101 (2012)
58. Kampstra, R., Ashayeri, J., Gattorna, J.: Realities of supply chain collaboration. Int. J. Logis. Manag. **17**, 312–330 (2006)
59. Huang, G., Lau, J., Mak, K.: The impacts of sharing production information on supply chain dynamics: a review of the literature. Int. J. Prod. Res. **41**, 1483–1517 (2003)
60. Kalcsics, J., Nickel, S., Puerto, J., Rodríguez-Chía, A.M.: Distribution systems design with role dependent objectives. Eur. J. Oper. Res. **202**, 491–501 (2010)
61. Bertazzi, L., Paletta, G., Speranza, M.G.: Deterministic order-up-to level policies in an inventory routing problem. Trans. Sci. **36**, 119–132 (2002)
62. Cao, D., Chen, M.: Capacitated plant selection in a decentralized manufacturing environment: a bilevel optimization approach. Eur. J. Oper. Res. **169**, 97–110 (2006)
63. Calvete, H.I., Galé, C., Oliveros, M.J.: Bilevel model for production-distribution planning solved by using ant colony optimization. Comp. Oper. Res. **38**, 320–327 (2011)

Determinants of Corruption

César Leonardo Guerrero-Luchtenberg[(✉)]

University of Zaragoza, Ciudad Escolar s/n, 44003 Zaragoza, Teruel, Spain
cleonardo@unizar.es

Abstract. In this paper we study determinants of the level of certain types corruption in a society. To that end, we apply the simplest technique used in Evolutionary Game Theory, namely, the replicator dynamics with two types of agents, corrupted, and not corrupted. Through a learning interpretation of that technique, we obtain the main determinants of corruption are the initial proportion of corrupted people, and the relative pecuniary gain of being corrupted, relative to the pecuniary gain of being not corrupted. The model applies to all types of corruption for which the larger the number of corrupted people is, the larger the expected payoff of being corrupted will be.

1   Introduction

The main objective of this paper is to find determinants of corruption in general, although we concentrate on what is called systemic corruption and micro-corruption (see O'Hara (2014)), that is, widespread corruption, and hence we concentrate in those types of corruption in which the more corrupted people there is in the society, the more beneficial to be corrupted is.

The importance of studying corruption in economics comes mainly from the fact that corruption and social capital are related issues, both of which are very important in order to promote growth and development. In the paper by Mauro (1995) it is shown that corruption is associated with low income, although it has not been established the direction of causation yet. Also, corruption has been related to political institutions and political outcomes (see Persson et al. (2003)), which in turn are crucial to growth, as argued by North (1991). For more on the negative correlation between corruption and growth, see Besley (2006) — Figure 1.6, among others—. Similarly, there is evidence that there exists a positive correlation between social capital and growth, as it is shown by Temple and Johnson (1998). They construct an index of social capital which turned out to be a good predictor of economic growth. One of the main ingredients of that index is again the level of corruption in the society. See also Fukuyama (1995) for more arguments in favor of the positive correlation between social capital, honest behaviour (not corruption) and growth.

In this paper, however, we strongly depart from standard studies on corruption, as most of them concentrate on public corruption by officials, governments, power elites, tax havens, abuse of power, etc. That is, for us corruption is not merely vested interests (of the most powerful people of the society) against the

© Springer International Publishing Switzerland 2016
R. León et al. (Eds.): MS 2016, LNBIP 254, pp. 104–111, 2016.
DOI: 10.1007/978-3-319-40506-3_11

common good (See O'hara (2014)). We will say that an action is corrupted, if that action violates or breaks the rules prevailing in the society. An agent is corrupted, then, if it commits a corrupted action. Therefore, in particular, all the informal economy is a form of corruption. This last form of corruption is not always denounced in the literature.[1] Other forms of not honest behaviour can be taken into account in our model as corruption, as for instance, at the university, students copying from one another, or even stealing exams from professors. These last types of dishonest behaviours may have also, of course, a higher pecuniary payoff than the honest behaviour, in expected future returns of course. A priory, we conjecture that for both tax evasion, and dishonest behaviour at the university, it is true that the larger the number of people willing to cheat is, the larger its expected payoff is, as the probabilities of being discovered would be lower when there is a lot of corrupted people in the society. So our model applies to these types of corruption.

Thus, our model also aims at describing when a worker is willing to cheat the firm in which it is working. For instance, when an unwatched worker is not working properly, precisely because not all workers can be watched. For instance, in a super market, when a worker is willing to steal food, alcohol, or even money if possible. On the other hand, workers might be also cheated by bosses, for instance, when they are not paying extra hours, though this behaviour is conceded or consented by workers, as otherwise they may lose the job. These types of corrupted behaviour seem to be exacerbated in crises times, as in current times in Spain for instance. Unfortunately, we have not found verifiable data (a common problem when trying to measure corruption, as workers would not denounced it, of course, for fear to lose the job), however news journals in Spain have been reporting that even more than 50 % of extra hours are not paid in current times. It is clear in these last examples, that the larger the number of corrupted workers is, the more difficult to watch them is.

We obtain then that the main determinants of corruption in the long run are the proportion of corrupted people there is in the society at the initial moment of analysis, and the relative average gain in pecuniary terms of being corrupted, relative to the average gain of all the population, corrupted and not corrupted. Having only two types, the payoff of a type, say not corrupted, is larger than the average, if and only if its payoff is larger than the payoff of the other type, say corrupted people. Therefore, the second determinant is the relative gain of being not corrupted, relative of the gain of being corrupted. Our interpretation is the following. A given agent, at a given moment of its life, is faced with the election of following the rules, or not. Then, we postulate that the given agent

---

[1] This form of tax evasion may account for huge quantities of value. For instance, if the informal sector is of 25 % of GNP, and taxes are on average 30 % or 40 % for small firms, the tax evasion may account for at least between 7.5 % and 10 % of the GNP of a country. Estimates tell that, for instance, the informal economy in Spain is more than 22 %, and in Italy or Greece are more than 25 %. For countries in Africa, for instance Zimbabwe, the informal economy is more than 59 % of GNP. See, for these numbers and many others, Schneider (2002).

will tend to cheat if it is profitable in pecuniary terms, and it will be profitable depending on how much corrupted people there is in the society at the given moment. If there are too little corrupted people, it is expected that to cheat is less profitable than not to cheat, and conversely, if there is a lot of corrupted people, it is expected to cheat is more profitable than not to cheat.

The rest of the paper is as follows. In Sect. 2 we roughly revise the related literature. Section 3 lays down the model and Sect. 4 presents the results. In Sect. 5 we discuss the results, and finally in Sect. 6 the conclusions are presented.

2   Related Literature

The closest studies to ours are clearly those which searched for the determinants of corruption. Perhaps the most recent and related to our study is given in Salih (2013). As it can be inferred from that last paper, most of the studies on the issue have concentrated on empirical relationships between corruption and the proposed determinant variables. Among these variables, as reported in Salih (2013), we have: (1) Income (negatively correlated); (2) Government size (negatively correlated); (3) Foreign Direct Investment (potential negative correlation); (4) Economic freedom (not clear sign of the correlation); (5) Quality of Judiciary system (negatively correlated); (6) Import share (negatively correlated); (7) Trade openness (negatively correlated); (8) Foreign aid (not clear sign of the correlation); (9) Inflation (positively correlated). In sharp contrast to these literature, we propose two theoretical determinants, both of which appear as we apply the replicator dynamics to the issue.

Some comments are in order. Observe that for most of the determinants above named, it is heavily suggested that in rich countries, corruption is lower than in poor countries. This indicates some contradictions in the issue, which is to say, most studies in corruption concentrate on the type of corruption powerful people exercises, but most powerful people, owners or managers of big firms for instance, are in rich countries presumably. To our understanding, the data suggests the following interpretation: In rich countries corruption stays mainly at high levels of the society, but lower income people of those rich countries, tend to be less corrupted, than those lower income people of poor countries. In poor countries, corruption would be present everywhere, low income people, and high income people.

Now we come to the model.

3   The Model

There are two types in the society $j \in \{C, NC\}$, where C denotes the corrupted type, and NC denotes the not corrupted type. Denote by $(\rho(t), 1 - \rho(t))$ the profile of types, where $\rho(t)$ is the proportion of people of type C at t, and a similar definition to $1 - \rho(t)$ applies. Time is continuous, and the horizon is infinite. We may think that at a given moment of time, there are a proportion of corrupted people, and some new people is born, in a way that the existing ones and the new born are learning the type they will adopt.[2] The type C is characterized by a function of the proportion of types C of the form $u_C : [0,1] \rightarrow \mathbb{R}_{++}$, a function which represent the average pecuniary payoff that a corrupted agent expects to gain, if at t there are $\rho(t)$ proportion of corrupted people. Similarly, the type NC is characterized by a function of the proportion of types C of the form $u_{NC} : [0,1] \rightarrow \mathbb{R}_{++}$, which represent the average pecuniary payoff that a not corrupted agent expects to gain, if at t there are $\rho(t)$ proportion of corrupted people. Observe that both pecuniary payoffs are strictly positive, which is assumed for simplicity (to avoid details when a denominator is zero). Observe that we use the word *average,* to emphasize the idea that it contains both, benefits and counter benefits of being corrupted, including the risk of going to jail, but measured in pecuniary terms. Moreover, as most probably not all forms of corruption have the same payoff, we prefer to fix the type of corruption, as it was insinuated in the Introduction.

Finally, we impose a replicator dynamics over the types of the model. For a general presentation of the replicator dynamics, see, for instance, Vega-Redondo (1996). To that end, let us define $\bar{u}(\rho(t)) = \rho(t)u_C(\rho(t)) + (1 - \rho(t))u_{NC}(\rho(t))$, which is the average pecuniary payoff in the population. Therefore, we postulate that $\rho(t)$ is determined by the following equation:

$$\frac{d(\rho(t))}{dt} = \dot{\rho}(t) = \rho(t)\frac{(u_C(\rho(t)) - \bar{u}(\rho(t)))}{\bar{u}(\rho(t))}.$$

That is, the proportion of corrupted people increases at t, if and only if, there are corrupted people in the society at t, and the pecuniary gain of corrupted people outnumber the average pecuniary gain. As we have said it, we do not impose a biological interpretation of the equation, that is, we do not postulate people are automata who acquire a genotype which determines its future behaviour. Instead, we propose a learning interpretation: People tend to imitate those who are the most successful in the society, so if to be corrupted generates higher payoff than average, people will tend to follow that behaviour and, vice versa, if people observe that to be not corrupted generates higher gains than average, people will tend to imitate that behaviour.

It is a simple exercise to see that the previous differential equation is equivalent to (Notice that if $\rho(t) = 1$ at some t, then we have $\dot{\rho} = 0$, that is, the proportion does not change at t, as it must be: A given type only can decrease if there exists another competing type. This reasoning is very pertinent in biology, however in our context this condition is a little bit stronger, as we may have, in principle, $\rho(t) = 1$ but $u_{NC}(1) > u_C(1)$. We will rule out that situation. In more general situations, we should be forced to consider mutations.):

$$\dot{\rho}(t) = \rho(t)(1 - \rho(t))\frac{(u_C(\rho(t)) - u_{NC}(\rho(t)))}{\bar{u}(\rho(t))}.$$

---

[2] Alternatively, we may assume a constant number of infinitely lived agents, who learn over time. The interpretation just proposed is, to our understanding, more realistic.

Therefore, provided there are some corrupted people in the society, the proportion of corrupted people increases, if and only if, the expected pecuniary gain of being corrupted is higher than the pecuniary gain of being not corrupted.

Now we come to the assumptions.

A1 $u_C : [0, 1] \to \mathbb{R}_{++}$ is increasing and $u_{NC} : [0, 1] \to \mathbb{R}_{++}$ is decreasing, and both functions are continuous.

A2 $u_{NC}(0) > u_C(0)$, and $u_{NC}(1) < u_C(1)$.

Both assumptions A1 and A2 would be satisfied if tax evasion is the type of corruption under study, as we commented before. We assume continuity just for simplicity of the exposition, as in this case we can apply standard theorems to prove existence of solutions to ordinary differential equations.

Some standard definitions follows. A steady state is a profile $(\rho^*, 1 - \rho^*)$ with $\rho^* = \rho(t^*)$ such that $\dot{\rho}(t^*) = 0$. This profile is said to be globally stable on $[0, 1]$, if for any $\rho(\tilde{t}) \in (0, 1)$, then $\rho(t)$ tends to ρ^*, as t tends to infinite. That is, the proportion of corrupted people converges to the state ρ^*. Similarly, this profile is said to be asymptotically stable, if there exists an ε (it may be arbitrarily small) such that for any $\rho(\tilde{t}) \in (\rho^* - \epsilon, \rho^* + \epsilon)$, then $\rho(t)$ tends to ρ^*, as t tends to infinite. Observe that both $\rho_1^* = 0$, and $\rho_2^* = 1$ are steady states of the system.

On the other hand, notice that given A1-A2, there exists a unique $0 < \rho^* < 1$ such that $u_C(\rho^*) - u_{NC}(\rho^*) = 0$. Thus, that ρ^* is the unique interior steady state of the system, if A1 and A2 are assumed.

Now we come to the main theorem of this paper.

4   Result

Theorem 1. *Assume A1 and A2, and the replicator dynamics on $\rho(t)$. Therefore, given the initial condition $\rho(\tilde{t})$, there exists a unique function $\rho(t)$ satisfying the replicator dynamics and that initial condition. Moreover, we have that: (a) If $\rho(\tilde{t}) < \rho^*$, then the unique solution of the equation, $\rho(t)$, tends to ρ_1^*; thus, in particular, $\rho_1^* = 0$ is asymptotically stable; (b) If $\rho(\tilde{t}) > \rho^*$, then the unique solution of the equation, $\rho(t)$, tends to $\rho_2^* = 1$, as t tends to infinite; thus, in particular, $\rho_2^* = 1$ is asymptotically stable.*

Proof. The proof is as follows. First, the existence of a unique solution satisfying the initial condition and the replicator dynamics follows at once from A1. This is a standard result in the issue. See, for instance, Arnold (1978), or any standard text book in the topic of ordinary differential equations. Now we come to the more specific results, for which it is sensible to prove it, once we know there exists a solution of the differential equation. Consider first Item (a). Notice that due to A1-A2, we have that $u_C(\rho) - u_{NC}(\rho) < 0$ for all $\rho < \rho^*$, and therefore, $\dot{\rho}(t) < 0$ for all $t \geq \tilde{t}$. Thus, $\rho(t)$ is decreasing and hence tends to $\rho_1^* = 0$ as t tends to infinite, as if $\rho(t)$ tended to a non-zero level, say $\bar{\rho} < 1$, $\rho(t)$ would continue decreasing, which would contradict that $\rho(t)$ has a limit as t tends to infinite.[3] The proof of (b) is analogous, and hence omitted. ∎

Some technical comments are in order. Notice that, assumed A1, A2 is a crucial assumption to obtain a unique interior steady state. Indeed, given A1, If we dropped A2, either ρ_1^* or ρ_2^* may become globally stable. If we required $u_{NC}(1) > u_C(1)$, for instance, ρ_2^* would be globally stable. And if we required $u_{NC}(0) < u_C(0)$, ρ_1^* would be globally stable. On the other hand, if we dropped the monotonicity requirements in A1, other interior steady states may appear, and therefore there may exist a society in which coexist corrupted and not corrupted people in the long run. Perhaps for some types of corruption that result might be the appropriate one.

5 Discussion

As said it, our objective is to find determinants of corruption. In the Theorem above, Item (a) states: if the proportion of not corrupted people is large enough at the initial moment of analysis, say $\tilde{t}$ (normally, $\tilde{t} = 0$), that is, if $\rho(\tilde{t}) < \rho^*$, and the pecuniary gain of being not corrupted is larger than the pecuniary gain of being corrupted ((A2) and $\rho(\tilde{t}) < \rho^*$), then in the long run there will not be corrupted people in the society. On the other hand, Item (b) states: if the proportion of corrupted people is large enough at the initial moment of analysis, that is, if $\rho(\tilde{t}) > \rho^*$, and the pecuniary gain of being corrupted is larger than the pecuniary gain of being not corrupted ((A2) and $\rho(\tilde{t}) > \rho^*$), then in the long run there will be only corrupted people in the society.

Finally, as commented in the previous section, if the monotonicity requirement in A1 was dropped, other interior steady states may appear, being some of them asymptotically stable. Therefore, other types of corruption may be studied under our set-up, by correcting the assumptions appropriately.

From our result it is possible to infer recommendations to governments, if the objective is to reduce corruption in the society. Indeed, given the proportion of corrupted people at a given moment of time, governments may strengthen punishments to minor faults, so that to lower the expected pecuniary gain of being corrupted. In the end, any strengthen of punishments may reduce corruption. However, if it is done to minor faults, this may imply punishments to juvenile people, and thus harsh punishments in the future for more serious faults become more credible. Definitely, one lesson of our paper is the following: If to be corrupted is profitable (due to weak punishments, for instance), we can hardly expect to have a non corrupted society.

Moreover, our approach heavily suggests that the problem of corruption in general, it is a problem of the whole society. The previous paragraph suggested a possible mechanism to fight against corruption implemented by governments

---

[3] A more rigorous argument is that having $\lim_{t \to \infty} \rho(t) = \bar{\rho}$, thus $\lim_{t \to \infty} \dot{\rho}(t) = 0$, and therefore $\bar{\rho}(1 - \bar{\rho})\frac{(u_C(\bar{\rho}) - u_{NC}(\bar{\rho}))}{\bar{u}(\bar{\rho})} = 0$. Hence, we must have $\bar{\rho} = 0$, as $\bar{\rho} < \rho^* < 1$.

which essentially places the issue on education and learning, learning from incentives. That learning would be much more effective if parents, from the very beginning of children life, are teaching to be honest, by example, by effectively lowering the expected payoff of being not honest, as the replicator dynamics suggests, under our interpretation.

This way, if the society is successful in lowering the expected payoff of corrupted people, firms and workers would be benefited if there is less micro-corruption in the society. Perhaps those corrupted agents that are not punished would have a very high return, however those that are punished, are severely punished.

Of course, our study suggests the question: Is it possible to reduce the type of corruption we study here without punishments? Perhaps the answer is yes, but our model does not provide the answer. The model is not saying that governments and parents must severely punish, but it is suggesting the reason why that behaviour may work. The crucial message is not to punish, but to lower the expected payoff of corrupted people, given the proportion of corrupted people there is in the society at the given moment of analysis.

6   Conclusions

In this paper we use the standard replicator dynamics in continuous time to find determinants of certain types of corruption. We find two determinants, one of them being the relative gain of being corrupted in relation to the gain of being not corrupted, and the other is the initial proportion of corrupted people.

This conclusion applies for certain types of corruption which satisfies our two assumptions. If those assumptions does not hold, then it may appear situations in which, depending on the initial proportion of corrupted people, the society may converge to an interior steady state in which there is both, corrupted and not corrupted people.

The type of corruption for which our model can be applied are those that normally causes much unrest in the society, both for firms and workers in particular. Indeed, a firm which needs not to spend resources to watch workers, will have higher benefits. Presumably, those benefits will go to workers as well, at least in part. Although one may strongly doubt about that, the data we have already commented here suggests that is the case. Indeed, that is what the negative correlation between corruption and income suggests, as in general rich countries (with a high average income per-capita), is where corruption is less present.

Therefore, the results depend on the type of corruption is considered. This begs the question: Is it possible to set general model in order to study all types of corruption? We leave for future research to answer that question. However, our intuition is that it may be possible, by constructing an appropriate unified corruption index.

References

Arnold, V.I.: Ordinary Differential Equations. MIT Press, Cambridge (1978)

Besley, T.: Principled Agents? The Political Economy of Good Government. Oxford University Press Inc., New York (2006)

Fukuyama, F.: Trust – The Social Virtues and the Creation of Prosperity. Free Press, New York (1995)

Mauro, P.: Corruption and growth. Q. J. Econ. **110**, 681–712 (1995)

O'Hara, P.A.: Political economy of systemic and micro-corruption throughout the world. J. Econ. Issues **XLVIII**(2), 279–307 (2014)

Schneider, F.: The Size and Development of the Shadow Economies of 22 transition countries and 21 OECD countries, IZA Discussion Paper No. 514 (2002)

North, D.C.: Institutions. J. Econ. Perspect. **5**, 97–112 (1991)

Persson, T., Tabellini, G.: The Economic Effects of Constitutions. MIT Press, Cambridge (2003)

Salih, M.A.R.: The determinants of economic corruption: a probabilistic approach. Adv. Manag. Appl. Econ. **3**(3), 155–169 (2013)

Temple, J., Johnson, P.: Social capability and economic growth. Q. J. Econ. **113**, 965–990 (1998)

Vega-Redondo, F.: Evolution, Games and Economic Behaviour. Oxford University Press, Oxford (1996)

The Design of the Logistics Operation of Food Distribution in Different Localities in the South of Bogotá Based on a Multistage Model

Oscar Herrera Ochoa[1(✉)], Oscar Mayorga[1],
and Angélica Santis Navarro[2]

[1] Universidad de La Salle, Bogotá, Colombia
{ojherrera, osmayorga}@unisalle.edu.co
[2] Universidad Cooperativa de Colombia, Bogotá, Colombia
angelica.santisn@ucc.edu.co

Abstract. Bogota has a problem with food supplies framed by the inefficiency in the capillary distribution, raising the costs and creating difficulties for food access in the social strata 1 and 2. Therefore, the city's mayor's office created the Masterplan for supply and food security (PMASAB), which includes the construction of logistic platforms located strategically in selected locations in order to reduce the food handling and transportation costs. Therefore, a multi-stage model for food distribution, that allows to improve the efficiency of the logistics operation in four marginal areas of the city with the use of three platforms, was designed. In the first stage, the load allocations are determined to be distributed from each platform based on mixed integer mathematical programming, the second develops a heuristic model to determine the vehicle routing and the third the routing simulation was performed, getting the flexibility for this operation strategies.

Keywords: Food · Heuristics · Multi-stage model · Mixed integer programming · Simulation

1 Introduction

The food supply in Bogotá D.C. is disorganized. Every day about 26.300 producers deliver foodstuffs to 140,000 different operators, there are on average 3 intermediaries (agents that do not add value, but if added cost) by chain, representing 21 % of the final price of the food [1], thus having an important negative effect on consumers, especially with the low-income and considering that Bogotá is a city characterized by a marked socio-economic inequality in which prevail strata 1 2 and 3; (84.5 %) [2]. For this reason, the city's mayor's office created the Masterplan for supply and food security (PMASAB), which compared to logistical issues, seeks to overcome the traditional intermediation in supply chains to achieve that the actors located at the ends of the chain feel to negotiate directly and to reduce inefficiencies of the current string, so that the consumer will be benefited by a remarkable reduction of prices. Facing this situation, a design of the logistics operation of food distribution in the localities of San

© Springer International Publishing Switzerland 2016
R. León et al. (Eds.): MS 2016, LNBIP 254, pp. 112–122, 2016.
DOI: 10.1007/978-3-319-40506-3_12

Cristobal Ciudad Bolivar, Bosa, Usme in the southern area of Bogotá D.C. were investigated. This work is aimed develop an approach through the use of a multistage model. A correlational analysis of information was used. The information was obtained from historical data and by direct observation in order to establish theoretical-practical judgments concerning the problem. This research was classified as descriptive correlational of field [3].

2   Literature Review

2.1   Models of Logistics Operation by Stages

Taking into account that a model is the simplified representation of reality by a mathematical system that helps to improve the level of understanding of its operation [4], and given that there are models with certain algorithms, complexity leads to a phased solution strategy, which consists of dividing the problem by parties and resolving separately each of these parts. This use has been found in the solution of problems of location of logistics platforms, with an initial phase of location of the central platform, then to select the means of transport and the number of trips to be performed [5]. Other applications seek to minimize the costs of inventory throughout the supply chain considering non-linear restrictions [6] and also problems of variability in inventory of cyclical demand security [7]. Studies of cost-sharing are in the supply chain where a stochastic mathematical model for analysis of networks of multiple stages, intends to [8]. For problems of interdependence of the different levels of the chain of supply orders is being formulated a mathematical model of the stochastic network for the analysis of the order and the distribution of costs [9].

2.2   Mathematical Model and Solution Method

The case of refill several clients with possibility of supply from platforms or deposits, the model that fits is the VRP (Vehicles Routing Problem) based on multi-deposits MDVRP. Most of the exact algorithms solving the VRP classical model are difficult to adapt to meet the MDVRP [10]. According to [11], these problems are the most common and easy to understand in order to develop formulations based on integer programming. Recently, in [12] is proposed mathematical formulations to meet various types of vehicle routing problems including the MDVRP and in [13] is presented a whole linear program to solve the MDVRP with heterogeneous fleet with time windows. At [14] proposes a model of mixed integer linear programming, MILP to minimize the cost of routing in the HFMDVRP, in which the heterogeneous fleet of vehicles are available. The above in [15] proposes a mixed integer linear programming model and a branch-and-bound procedure for the MDVRP fleet fixed pick-up and delivery. Other settings of the MDVRP development of exact algorithms are exposed in [16] where the MDVRP is studied, including limitations on the length of routes. In [17] is shown the MILP (Mixed Integer Linear Programming) model and linear programming whole mixed in the problem in which the heterogeneous fleet of vehicles available and the maximization of total net revenues as an objective function, while restrictions on maximum and minimum demands are set. On the other hand, heuristic methods are used to solve the problems of optimization through an intuitive approach according to the structure of the problem [18], where is considered a limited search space exploration and which gives acceptable solutions at the time of calculation. Many of these heuristics can be extended to handle additional restrictions to the VRP [19]. In [20] explores the problem of collection and delivery without inclusion of multiple deposits in an integrated fashion (MDVRPPD). [21] proposed a heuristic together with the algorithm branch and price for the minimization of costs in heterogeneous fleet MDVRP. On the other hand in [22] proposes a three-stage heuristic. First a pre-processing for node grouping, then a formulation of the problem with more compact MILP, then a cluster for their development. Also in [23] developed a heuristic based on the integer programming. [19, 24–26] study the most commonly used heuristics: algorithm for nearest neighbor, algorithm of savings, insertion methods, methods assign first - route later, the first route - assign later, petals, and procedures of local search algorithm.

3   Methods

Developed applied research is technology to develop a non-cross-correlation type experimental research design, based on historical data, by direct observation, and other collected data from secondary sources. Also previously validated [27] has worked with theoretical foundations, engaged in the description of different variables at a given time to determine its correlation or causal [28] relations and to propose improvement of the performance of the unit of analysis and to make it more efficient. Methodological stages are summarized in Fig. 1, where each one by itself generates practical and theoretical research contributions and where each procedure used from operations research and statistics have epistemological basis, that transcends the practicality and utility level merely technical [29], taken from [30]. Facing to the type of information that depends on variables which are the most relevant in the operation of supply, collection and subsequent analysis, the characteristics are shown in Table 1.

Methodological phases can be seen in Fig. 1 that follows:

Table 1. Source, treatment and use of the information by stage

Data type	Way of obtaining	Mean or tool used	Stage
Distances between clients (nodes) Time between clients (nodes)	Web application	Google Maps, geo positioning systems satellite (G.P.S.)	1-2-3
Demand Offer	Specialized information centers: City Hall Mayor of Bogota and District Planning Secretariat	Libraries and Web page	1-2

(*Continued*)

Table 1. (*Continued*)

Data type	Way of obtaining	Mean or tool used	Stage
Fixed and variable costs	Specialized information centers: Corabastos	Newsletters from monthly prices. Direct interview	1
Time of upload and download	Head of operations of Corabastos Research Group Ingenio Induspymes (University Cooperative of Colombia)	Research report	2-3
Allocation of loads from platforms to centers tradesmen	Computer application	Solver of EXCEL	1-2
Routes of vehicles for the remaining products	Computer application	Programming in EXCEL	2
Statistical behavior of variables.	ProQuest databases	Web	3
Transfer and return times	Computer application	EXCEL®	
Statistical behavior of time of upload and download	Computer application	EXCEL StatGraphics StatFit	2-3
Operation of routing strategies	Computer application	Promodel	3

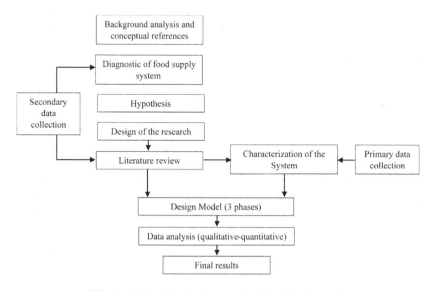

Fig. 1. Methodological stages developed in the work

4 Mathematical Modeling and Solution Proposal

Given the approach to the improvement of the access of food at fair price in the city of Bogotá, the supply operation contemplating the construction of logistic platforms located strategically in different locations in the South of the city was designed. The food arrive and will be distributed to centers of shopkeepers (CT) at each locality to be sold directly to consumers. In this work sets the design of the logistics operation of food distribution in the localities of San Cristóbal, Ciudad Bolívar, Bosa and Usme of Bogotá D.C., through a model developed by stages to simplify your design problem. This model would allow that there is less handling of the food (food security) and reduces the transporting and intermediation cost, thus the prices for the final consumer are lower.

4.1 General Description of the System and the Model

Food distribution would be made from the platforms: Lucero, Bosa and Usme, to each of 27 centers of shopkeepers (CT) distributed in four localities of study. Considering the location, capacity, coverage, and demand for these. Accordingly, the model should allow supply to each CT is performed by a single platform (although this is not it your locality and not to exceed its capacity). Thus, the developed model includes three stages: in the first full load allocations is determined by product families to distribute from each platform based on mixed integer mathematical programming. This information makes it possible to programming for distribution with full loads to each CT for three food chains established (Fruver, Potato and grocery). The second develops a heuristic model to determine the routing of vehicles for the distribution of partial loads that remained from the three chains. Where is made the transport of food by mixing all products and the third is the simulation of routing set forth above in order to assess its dynamics and thus define different actions to develop the operation in the most convenient way possible.

4.2 Mathematical Approach First Stage

The following model is built from the conditions of operation of the system in question and supported in formulations of various authors. Where the model represents the transport of products in a network with "m" sources and "n" destinations according to [31]. Within the variable do not provide the type of product as subindice, because the model is ran by type of food chain because conditions differentiated from platform capabilities for each of the chains and vehicles by locality. The objective is not to make complex the mathematical model. The objective function (1) includes both variable costs and fixed costs, also based in [19, 32] is defined (2) with a design of distribution routes to multiple clients using various sources of the system, defining in this way a single source by destination. Supported by [33, 34] is set (3) looking for the minimization of fixed costs, having a constraint of fixed maximum cost for the mobility of vehicles. Equation (4) is the restriction of supply defined in the model of conventional transport [31]. Taking into account multiple deposits, the demand equation is set to (5). On the other hand, provides an unrestricted non-negative variable and another binary, making part of the mixed integer programming [13, 35]. The specific structure of the proposed model is then set:

Decision variables and parameters:

X_{ij} = Number of units to transport of the same type, from the logistics Platform i, to the consumer or Customer Center j within the supply system nodal network.

Y_{ij} = binary variable that is activated if it takes the value of 1, or deactivates if it takes the value of 0, the use of the logistics Platform- i, downtown shopkeeper or customer j within the system nodal network.

C_{ij} = Cost unit of transport of the product, supply or logistics Platform i, Centre shopkeepers j supply, nodal network which is given by the average cost of purchase of the product in the market.

CF_{ij} = Cost fixed transport logistics Platform i, Centre shopkeepers j supply, nodal network given by the projection of the vehicle resource monthly rental.

$Cmax_{ij}$ = Maximum cost allowed by the transport from the logistics Platform i, Centre shopkeepers j within the supply system nodal network.

a_i = capacity available for the logistics Platform i supply.

b_j = request of demand from the center of shopkeepers j

Mathematical structure:

$$\text{Min } Z : \sum_{i=1}^{m} \sum_{j=1}^{n} C_{ij}X_{ij} + \sum_{i=1}^{m} \sum_{j=1}^{n} CF_{ij}Y_{ij} \text{ Objective function} \tag{1}$$

Subject to

$$\sum_{i=1}^{m} Y_{ij} = 1 \; \forall \; j = 1, \ldots, n \text{ Connectivity restriction} \tag{2}$$

$$\sum_{j=1}^{n} CF_{ij}Y_{ij} = C\max \; \forall \; i = 1, \ldots, m \text{ max cost constraint allowed} \tag{3}$$

$$\sum_{j=1}^{n} X_{ij} \leq a_i \; \forall \; i = 1, \ldots, m \text{ Restriction of supply or capacity} \tag{4}$$

$$X_{ij} - Y_{ij} * b_i = 0 \; \forall \; i = 1, \ldots, m \text{ y } j = 1, \ldots n \text{ restriction of demand} \tag{5}$$

$$X_{ij} \geq 0 \text{ Restriction of no negativity} \tag{6}$$

$$Y_{ij} \text{ Binary} \tag{7}$$

4.3 Approach Second Stage

Using a heuristic method is more flexible than the exact methods, which allows to incorporate terms that are difficult to model [25], such as curfews, vehicular accessibility on logistics platforms and the vehicle capacities. Contemplating various heuristics, was defined as the best option the use of heuristics of savings or Clarke and Wright parallel version in an asymmetric network, that allowing better time in its development. Considering the allocation of routes generally for the entire network through savings in the distances, to then include windows of time or the inherent restrictions of the system. At this stage, the routing of vehicles is defined for products remaining to distribute after the sent vehicles to loaded 100 % of its capacity, combining products of three food chains, where the use of the vehicle is given below its maximum capacity.

4.4 Approach Third Stage

Once defined routing through heuristics, the simulation it's determined in the applicative Promodel. For this, first defined the inherent and most important system variables (transfer times, demand for downtown shopkeeper and time of upload and download), then determines their statistical behavior and later the construction of the simulation model is carried out by logistics platform. The times of transport vehicles and load variables were developed a normal distribution, through the central limit theorem [36], its calculation was based on this premise, since there were no information by being a system that so far is intended to establish. The loading and unloading times variables, was defined as dependent on the load to be transported according to a linear function positive.

Cochran's formula and the modified Tchebycheff's theorem was used. 20 runs or replies as the value to validate and to work with the simulation model were resulted. After validating the model through historical data comparing variables of load to be transported with loading and unloading times and finally the times of operation of supply, the experimentations of the model. The specific actions of the experimentation were:

1. Evaluate the option of using another vehicle for the routing of the Lucero platform, since the current operation time is at 9.01 h and exceed 8 h of daily work.

2. Take into account changes in the behavior of the transfer times, anticipation of unpredictable or unexpected delays.

3. See a technology that handle the load vehicles on platforms.

5 Results

As referenced above, the general model was designed to establish the logistics operation of food supply in the localities of Ciudad Bolívar, Bosa, Usme and San Cristobal in Bogota taking into account chains Fruver, Pope and groceries. For the distribution to 27 centers of shopkeepers CT located by each Zonal Planning Unit (ZPU) in these areas. It was developed in the first stage, 15 runs of the optimization model in Solver of

Table 2. Average cost for food chain a month

	Grocery *	Fruver *	Pope *
Minimum	$2,025	$1.182	$482
Average	$2.117	$1.233	$496
Maximum	$2,174	$1,278	$511
Variance	$1.688	$764	$75
* values in millions of pesos

Excel and the demand was changing according to normalized data [38]. It was done during half-month in order to evaluate their behavior at least 50 % of the planning horizon; and thus the full month supply total costs were obtained as shown in Table 2.

On the other hand, it have specifically the following mappings: platform of the Lucero caters to 9 CT equivalent to 100 % of their own locality, in addition 1 CT of the locality of Bosa (ZPU Apogeo), 2 CT from the locality of San Cristobal (ZPUs Sosiego and 20 July) and 4 CT of the locality of Usme (ZPUs Comuneros, Danubio, Usme city and great Yomasa). Bosa platform caters to 4 of 5 centers tradesmen CT 80 % is local and Usme platform caters to 3 of 7 centers tradesmen CT equal to the 48.8 % of his locality, also 4 CT from the town of San Cristobal (ZPUs San Blas, La Gloria, Libertadores and Rural). Based on the results of the previous optimization model and taking into account the quantities of products of the three chains that are remaining for distributing, it developed the heuristics of savings (second stage) in laying down the routes to cover from each platform to meet the remaining demand which was not supplemented with the assignment of vehicles loaded to 100 %, assessed by each pair of points of greater savings, three route options where the restrictions of capacity and delivery time is met, obtaining routes for each of the platforms as well: Lucero, 15 possibilities were evaluated (each with 3 options). A total of 45 possible routes, of which 10 definitive route were obtained. From the Usme's Platform 3 possibilities were evaluated. A total of 9 possible routes, and 3 definitive routes were obtained, and from the Bosa's platform an (1) possibility with 2 options. The final routes by platform are in Table 3.

Table 3. Definitive routes

Lucero's Route	1	2	3	4	5	6	7	8	9	10
Tours of the CT	0,26,24,0	0,18,0	0,19,0	0,3,2,9,0	0,23,0	0,7,0	0,6,0	0,27,0	0,4,0	0,5,8,0

Usme's Route	1	2	3	Bosa's Route	1
Tours of the CT	0,22,25,0	0,17,0	0,15,16,0	Tours of the CT 0,14,12,11,0	

For stage 3, the operation routing set in the previous stage in Promodel was simulated. Resulting in that the time of operation of all activity on the platform the Lucero is 9.01 h, using 2 vehicles with a capacity from 5200 kg, and carrying out 10 routes defined. For the platform of Bosa, is the operation time is 2.80 h, using 1 vehicle from 9000 kg capacity and carrying out the defined path: 0,14,12,11,0; and finally for the platform of Usme, operation time from all activities is 6.01 h, using 1 vehicle of 5200 kg capacity and carrying out the 3 established routes. This done, three experiments are included as mentioned in item 4.4 on the simulation of the system for the platform of the Lucero (Ciudad Bolívar), because this is the platform that presents more saturation. For the first experiment, the total operating time decreases to 6.97 h using 3 vehicles. For the second experiment, the total operation time gives 9.25 h considering that 1 vehicle increases its operating time by 9.3 % and the transport by 13.4 %, and 2 vehicle increases its time in 1.3 % and 5.6 % for the operation time and transport respectively. For the experiment 3, total operation system time is reduced to 8.24 h using 2 vehicles. There is a reduction of 13 % in the time for each vehicle entrance to the platform and the 21.8 % over the total time of the distribution activity corresponding to 28 min. The use of means of transport included the 5200 kg capacity for all experiments.

6 Conclusions

The design of the logistics operation of supply through a model developed by stages, allows you to subdivide macro activities of supply for the localities of Bosa, Ciudad Bolívar, Usme and San Cristobal in less complex parts for analysis. This form of analysis of the system provides the general activities of the production management such as: planning, programming, and control. Each one of the stages is focused on these activities, which provides a good quantitative tool for decision-making and for the analysis of systems. On the other hand, the design of the logistics operation of supply of food through platforms, resulting in a greater for logistics operators results in more efficient supply, since established routes, schedules, detailed amounts, times, among other things, that will make effective decisions for timing of supply at low cost, while contemplating the conditions of evaluation of the system. Specifically, it was obtained that the allocation of customers or centers tradesmen to the platform the constellations of the locality of ciudad Bolívar was the largest with 16 assignments of possible 27 in four areas of study, representing 59.2 %. This essentially given to the fixed costs of operation of the fleet vehicles in Bosa are higher (82 %) compared to the cost of the platform Lucero in the locality of Ciudad Bolivar. The reason for this is that in the locality of Bosa 9 tons, are used while in Ciudad Bolivar are 5.2 tons. The locality of Bosa platform had an allocation of 4 centers of shopkeepers being the fewest assigned and taking a percentage of utilization capacity of 53.3 %, while the platform of Usme had less use of their capacity with the 26.6 %. This is because the requirement by Center of shopkeeper in Bosa is higher; on average 2.3 times greater than in the town of Usme due to higher population density per unit of UPZ Zonal planning. In this way, there is an underutilization of Usme platform, which could suggest that their future construction provides for one size lower than that established in this study, in such a way that your investment is less and can leverage best resources into technological elements for efficient operation. As obtained in the simulation, it says that the time dedicated by centers tradesmen CT the reception and unloading of vehicles is very short, approximately 7 % of the total activity of the routing, which is due to the activity of routing for loads combined with loading of vehicles is only referred and which will not 100 % of their capacity. In evaluated Promodel the results in terms of capacity, utilization, and time, which are consistent with the results of the proposed mathematical model.

References

1. Unidad Ejecutiva de Servicios Públicos UESP, Documento soporte técnico del Plan maestro de abastecimiento de alimentos y seguridad alimentaria de Bogotá, Unidad Ejecutiva de Servicios Públicos UESP, Bogotá (2005)
2. Secretaría Distrital de Planeación de Bogotá, Inventario de Información en Materia Estadística sobre Bogotá, Secretaria Distrital de Planeación, Bogotá (2011)
3. Behar Rivero, D.S.: Introducción a la Metodología de la Investigación. Editorial Shalom, Buenos Aires (2008)
4. Y. Estrada Perea, Modelación de la Distribución del Transporte de carga por carretera de Productos Colombianos, Universidad Nacional de Colombia, Medellín, Antioquia: Universidad Nacional de Colombia sede Medellín (2008)
5. Kalenatic, D., López Bello, C.A., González Rodríguez, L.J., Rueda Velasco, F.J.: Modelo para la localización de una plataforma de cross docking en el contexto de logística focalizada. Ingeniería 13(2), 36–44 (2008)
6. Liu, L., Xiaoming, L., Yao, D.D.: Analysis and optimization of a multistage inventory-queue system. Manag. Sci. 50, 365–380 (2004)
7. Gutiérrez, V., Vidal, C.J.: Inventory management models in supply chains: a literature review. Rev. Fac. Ing. Univ. Antioquia 43, 134–149 (2008)
8. Li, C.: An analytical method for cost analysis in multi-stage supply chains: a stochastic network model approach. Math. Model. 38, 2819–2836 (2014)
9. Li, C., Liu, S.: A stochastic network model for ordering analysis in multi-stage supply chain systems. Simul. Model. Pract. Theory 22, 92–108 (2012)
10. Montoya Torres, J.R., López Franco, J., Nieto Isaza, S., Felizzola Jiménez, H., Herazo Padilla, N.: A literature review on the vehicle routing problem with multiple depots. Comput. Ind. Eng. 79, 115–129 (2015)
11. Kulkarni, R., Bhave, P.: Integer programming formulations of vehicle routing problems. Eur. J. Oper. Res. 20(1), 58–67 (1985)
12. Baldacci, R., Mingozzi, A.: A unified exact method for solving different classes of vehicle routing problems. Math. Program. 120(2), 347–380 (2009)
13. Nieto Isaza, S., López Franco, J., Herazo Padilla, N.: Desarrollo y Codificación de un Modelo Matemático para la Optimización de un Problema de Ruteo de Vehículos con Múltiples Depósitos In: Proceedings of the 10th Latin American and Caribbean Conference for Engineering and Technology, Panama City (2012)
14. Dondo, R., Méndez, C.A., Cerdá, J.: An optimal approach to the multiple-depot heterogeneous vehicle routing problem with time window and capacity constraints. Lat. Am. Appl. Res. 33(2), 129–134 (2003)
15. Kek, G., Cheu, L.R., Meng, Q.: Distance-constrained capacitated vehicle routing problems with. Math. Comput. Model. 47, 140–152 (2008)
16. Contardo, C., Martinelli, R.: A new exact algorithm for the multi-depot vehicle routing problem under capacity and route length constraints. Discrete Optim. 12, 129–146 (2014)

17. Cornillier, F., Boctor, F., Renaud, J.: Heuristics for the multi-depot petrol station replenishment problem with time windows. Eur. J. Oper. Res. **220**, 361–369 (2012)
18. Díaz, F., Glover, Chaziri, H.: Optimización Heurística y Redes Neuronales. Paraninfo, Madrid (1996)
19. Olivera, Heurísticas para Problemas de Ruteo de Vehículos, Montevideo: Instituto de Computación, Facultad de Ingeniería, Universidad de la República (2004)
20. Nagy, G., Salhi, S.: Heuristic algorithms for single and multiple depot vehicle routing problems with pickups and deliveries. Eur. J. Oper. Res. **162**, 126–141 (2005)
21. Irnich, S.: A multi-depot pickup and delivery problem with a single hub and heterogeneous vehicles. Eur. J. Oper. Res. **122**, 310–328 (2000)
22. Dondo, R.: A cluster-based optimization approach for the multi-depot heterogeneous fleet vehicle routing problem with time windows. Eur. J. Oper. Res. **176**(3), 1478–1507 (2007)
23. Gulczynski, D., Golden, B., Wasil, E.: The multi-depot split delivery vehicle routing problem: An integer programming-based heuristic, new test problems, and computational results. Comput. Ind. Eng. **61**, 794–804 (2011)
24. Cordeau, J.F., Laporte, G., Savelsbergh M.W., Vigo, D.: Vehicle Routing In: Barnhart, C., Laporte, G. (eds.) (2007)
25. Contreras Pinto, M., Díaz Delgado, M.F.: Metodos Heuristicos para la solución de Problemas de Ruteo de Vehiculos con capacidad Universidad Industrial de Santander UIS, Bucaramanga (2010)
26. Díaz Parra, O., Ruiz Vanoye, J.A., Bernábe Loranca, B., Fuentes Penna, A., Barrera Cámara, R.A.: A survey of transportation problems. J. Appl. Math. **2014**, 1–17 (2014)
27. Vargas Cordero, Z.R.: La investigación aplicada: una forma de conocer las realidades con evidencias científicas. Educación **33**(1), 155–165 (2009)
28. Hernandez Sampieri, R., Fernandez Collado C., Baptista Lucio, M.d.P.: Metodología de la Investigación, 5ta Edición ed. McGraw Hill, Mexico D.F (2010)
29. Guadarrama, P.: Dirección y asesoría de la investigación científica. Magisterio, Bogotá (2009)
30. Gonzalez, C., Jaimes, W.A., Orjuela, J.A.: Stochastic mathematical model for vehicle routing problem in collecting perishable products. DYNA **82**(189), 199–206 (2015)
31. Winston, L.W.: Investigación de Operaciones Aplicaciones y Algortimos, 4th edn, p. 1418. Internacional Thomson Editores S.A, Mexico D.F (2005)
32. Kumar, S.N., Panneerselvam, R.: A Survey on the vehicle routing problem and its variants. Intell. Inf. Manag. **4**, 66–74 (2012)
33. Díaz, Parra, O., Cruz Chavez, M.A.: El Problema del Transporte, Centro de Investigación en Ingeniería y Ciencias Aplicadas, Cuernavaca, Morelos (2008)
34. Chopra, S., Meindl, P.: Administración de la Cadena de Suministro. Pearson, Nacaupal de Juárez (2008)
35. Montoya Torres, J.R.: Planeación del transporte y enrutamiento de vehículos en sistemas de producción. Rev. Cient. Ing. Desarrollo **13**, 85–97 (2003)
36. Maneiro, N., Mejías, A.: Estadistica para Ingeniería, una herramienta para la gestión de la calidad, Valencia, Venezuela: Dirección de Medios y Publicaciones Universidad de Carabobo (2010)
37. Alvarado, H., Batanero, C.: Meaning of Central Limit Theorem in University Statistics and Probability Textbooks, Estudios Pedagógicos, pp. 7–28 (2008)

Planning Freight Delivery Routes in Mountainous Regions

Carlos L. Quintero-Araujo[1,2(✉)], Adela Pagès-Bernaus[1],
Angel A. Juan[1], Oriol Travesset-Baro[3], and Nicolas Jozefowiez[4]

[1] Computer Science Department, Open University of Catalonia - IN3,
PMT Castelldefels, Castelldefels, Spain
{cquinteroa,ajuanp,apagesb}@uoc.edu
[2] Universidad de La Sabana, Chía, Colombia
carlosqa@unisabana.edu.co
[3] Observatori de la Sostenibilitat d'Andorra (OBSA),
Sant Julià de Lòria, Andorra
otravesset@obsa.ad
[4] Univ de Toulouse, INSA, LAAS, 31400 Toulouse, France
njozefow@laas.fr

Abstract. The planning of delivery routes in mountainous areas should pay attention to the fact that certain types of vehicles (such as large trucks) may be unable to reach some customers. The use of heterogeneous fleet is then a must. Moreover, the costs of a given route may be very different depending on the sense taken. The site-dependent capacitated vehicle routing problem with heterogeneous fleet and asymmetric costs is solved with the successive approximations method. The solution methodology proposed is tested on a set of benchmark instances. Preliminary tests carried out show the benefits, in terms of total costs, when using a heterogeneous fleet. In both cases, with and without site dependency, the increase in distance-based costs is mitigated by the use of heterogeneous fleet.

Keywords: Heterogeneous Site-Dependent VRP · Successive approximations method · Clarke-and-Wright Savings algorithm

1 Introduction

The delivery of goods plays an important role within the logistics and transportation sector. With the increase of demand, promoted by the growth of e-commerce or the development of new logistic strategies such as just-in-time, the number of batches to deliver also grows. Therefore, an accurate planning of the delivery routes is extremely important.

Road transport constitutes a major activity especially in urban areas, but also in many regions with limited access to other modes of transport. Globally, road transport is responsible for around a quarter of the EU's energy consumption and about a fifth of its CO_2 emissions. Moreover, freight delivery within cities is seen as a disturbing factor by its inhabitants. Freight related activities increase traffic congestion and impact on the

R. León et al. (Eds.): MS 2016, LNBIP 254, pp. 123–132, 2016.
DOI: 10.1007/978-3-319-40506-3_13

overall quality of life. Although all these inconveniences, goods need to be delivered and it is a major source of employment.

In this paper we focus on the planning of routes for the delivery of goods in mountainous regions. This problem applied to a general setting is very well characterized within the Operations Research/Management Science community by the Vehicle Routing Problem (VRP). The simplest version of the vehicle routing problem minimizes the total cost (usually based on distance) of visiting once all the customers departing from a depot. If the customers require some demand and the vehicle performing the route has limited capacity, then the problem is extended to the Capacitated Vehicle Routing Problem (CVRP). Another assumption is made regarding the type of vehicles available. The plain CVRP assumes that all the vehicles have the same capacity. On the contrary, if vehicles have different capacities the heterogeneous fleet CVRP is solved.

Mountainous regions may have special characteristics related to the topography. Some of the customers may be accessible only by regional roads, or after crossing mountain passes which in winter times may require that vehicles are equipped adequately. City centers, which usually have particular characteristics such as narrow streets and limited parking areas, may have even harder driving conditions (such as streets with slopes). All these characteristics limit the type of vehicles that can access certain areas. In particular, large trucks may be unable to access downtown areas or may experience difficulties in driving up and down the mountains. In such situation smaller vehicles seem more appropriate to be used to serve such customers.

The CVRP with heterogeneous fleet is then extended in order to incorporate the condition that some customers cannot be served by a particular type of vehicle. The particular characteristic of this application is that most of the customers are accessible for any type of vehicle except a reduced subset. This defines the Site-Dependent Vehicle Routing Problem, SDVRP.

The vast majority of the literature dedicated to solve VRPs assumes a symmetric cost matrix, thus neglecting the cost differences associated to drive in one direction or the other. It is clear that in mountainous regions, there can be large differences in cost if the route is uphill or downhill. The variant considered in this paper looks for routing all the nodes using an heterogeneous fleet within an oriented network (asymmetric costs) and ensuring that customers and delivery vehicles are compatible (site-dependent), HSDA-VRP (Heterogeneous Site-Dependent with Asymmetric costs VRP).

The contributions and aims of this paper are: (1) to present a heuristic procedure to solve the presented problem, and (2) to analyze how solutions change when the type of customers conditions the type of vehicles that can be used.

Regarding the solution procedure, we are interested in taking advantage of the existing algorithms for solving simpler versions of the VRP. Most of the metaheuristics used for solving enhanced versions of routing problems rely on efficient methods designed for solving classical VRPs. We also base our solution procedure on enhanced metaheuristics based on the classical Clarke-and-Wright Savings heuristic. A multi-round approach is presented similar to that in [1]. At each round a CVRP is solved considering a particular vehicle and assuming an unlimited number of vehicles of this type. Solutions are built from partial solutions found by solving CVRPs ensuring that

customers are served only with appropriated vehicles. The methodology used is efficient yet it is simple to implement and it has few parameters to adjust.

The rest of the paper is organized as follows. Section 2 reviews the vehicle routing
problem and its extensions, and the solution approaches presented in the literature. It
follows Sect. 3 with the description of the problem. In Sect. 4 we present our choice for
solving this problem. Section 5 presents the benchmark instances used to test the model
and compares different solution configurations. To conclude, the main findings are
enumerated together with future extensions to this work.

2 Literature Review

In this section, we first provide a general picture of the VRP and its variants and later
we discuss some of the solution approaches.

2.1 The Vehicle Routing Problem and Its Variants

The Vehicle Routing Problem is a classical Combinatorial Optimization Problem
(COP) present in many applications. The most popular VRP variant is the Capacitated
Vehicle Routing Problem. Given a fleet of identical vehicles located at a depot, the goal
is to provide routes to supply a set of customers with known demand so the total
delivery cost is minimized. Each customer can be visited only once and each vehicle
covers only one route. The total demand served by a particular route cannot exceed the
vehicle capacity.

The VRP has been extended in many directions in order to incorporate more
realistic characteristics. Regarding the properties of the fleet, this can be *homogeneous*
(all vehicles have the same characteristics) or *heterogeneous* (include a combination of
vehicles with different capacities), vehicles can have a limited driving range [2] or
drivers have scheduled breaks [3]. The network structure may be defined based on
symmetric costs (usually Euclidean distances are used) or *asymmetric* when other
factors such as fuel consumption or time used define the cost structure [4]. The routes
may be *open* (when vehicles do not have to go back to the depot) or *closed* (all routes
start and end at the depot, which is the most common assumption). In the multi-depot
VRP, customers are served from one of the several depots. Regarding the quality of
service, there can be conditions on the time when the customers can be visited, *i.e.*
time-windows are defined for each customer (VRPTW), or the customers are not only
receivers of demand but there is *pick-up and delivery* (VRPPD). In the *site-dependent*
VRP (SDVRP), a subset of customers is incompatible with some vehicles of the fleet.
For an extensive review of these and other VRP variants and applications, the reader is
referred to [5–10].

2.2 Solution Approach

Several approaches have been proposed to solve vehicle routing problems, both exact
methods and heuristics. Regarding exact methods, Kallehauge [11] reviews several

formulations of the VRP and exact solution methods. Baldacci *et al.* [12] provide different formulations of two classes of VRP variants. The set partitioning formulation shows to be the most appropriated when exact methods are used which combined with column-generation methods solve to optimality instances of up to one hundred customers.

Since the SDVRP belongs to the NP-hard category of problems [13], heuristic algorithms are usually used in practice. Many metaheuristics have been proposed to solve the many variants of the VRP, such as Tabu Search, Variable Neighborhood Search, Scatter Search or Genetic Algorithms. The description of site-dependent VRPs was first proposed in [14] and presented several simple heuristics to solve them. Chao *et al.* [15] developed a Tabu Search heuristic enhanced with deterministic annealing to widen the search space. Cordeau *et al.* [13] derived the SDVRP as a special case of the periodic VRP and solved it with a Tabu Search heuristic for the periodic VRP. Pisinger *et al.* [16] presented a unified heuristic based on the Adaptive Large Neighborhood Search framework capable to solve five variants of the VRP, being the site-dependent VRP one of them. To the best of our knowledge, there are few contributions for solving the VRP combining asymmetric costs with heterogeneous fleet and site-dependent characteristics. Recently, Yusuf [17] has presented a three-phase heuristic for solving the routing of ships considering two objectives, accessibility and profitability, as a multi-depot heterogeneous site-dependent VRP with asymmetric costs. The method proposed falls into the 'cluster first and route second' category.

3   Description of the Problem

The HSDA-VRP is a natural extension of the classical capacitated VRP. Formally, the HSDA-VRP is defined over a complete graph $G = (N, A)$ where $N = \{0, 1, \ldots, n\}$ is the set of nodes where node 0 is assumed to be the depot and the rest represent the customers that need to be visited. Each customer has a (deterministic) demand d_i. The set $A = \{(i,j) : i, j \in N, i \neq j\}$ contains the arcs that represent the road network. The set A considers that any pair of nodes is connected, and if this differs from the reality a large cost will be associated to it. The cost of travelling on the arc (i, j) is represented by c_{ij}. In the heterogeneous version, the cost of going from i to j is different from the cost of going from j to i. This cost can be related to distance, travelling time, fuel costs or other measures depending on the interest of the application.

Customer demands are carried by one of the vehicles available in the fleet. The set F includes all available vehicles. Each vehicle k will have particular characteristics. Most notable, the total cost of travelling from i to j will not only depend on the cost of the arc but also depends on the type of vehicle used. The cost of the arc is multiplied by a factor that depends on the vehicle type, v^k, (larger vehicles have higher variable cost than smaller ones), being then the total cost of the arc a three-index parameter, $c_{ij}^k = v^k \times c_{ij}$. Moreover, each route includes a fixed cost for using a vehicle, f^k. Therefore, the cost of a route corresponds to the sum of the fixed and variable costs of the arcs belonging to the route. Parameter Q^k denotes the maximum load that vehicle k can carry. Vehicles can serve only compatible customers, C^k, where $C^k \subseteq N \setminus \{0\}$ is the set of nodes that vehicle k can reach.

The AHSD-VRP goal is to find the routes that will serve all customers demand and minimize the total cost of travel. This problem can easily accommodate the main characteristics present when planning routes for freight transportation by road in mountainous regions: some type of vehicles may not be able to serve a subset of particular clients (thus requiring a fleet with multiple types of vehicles) and travelling costs depend on the direction of the route and type of vehicle. For a complete mathematical formulation of this problem, the reader should refer to [1].

4 Successive Approximations Method Algorithm

The solution approach that we propose is based on the Successive Approximations Method (SAM). This scheme is based on the ideas presented in [1]. The SAM algorithm is a multi-round process in which a solution is constructed in several steps. The number of rounds is limited by the number of type of vehicles. At each round a vehicle type is selected and routes are built to serve the subset of compatible nodes. When more routes than available vehicles are scheduled, a subset of routes is randomly discarded so no more vehicles than the available ones are used. Such routes are saved as a partial solution, and the nodes belonging to the discarded routes are extracted so they can be routed in the next round. In successive rounds, another type of vehicle is selected and a new VRP is solved assuming an unlimited number of vehicles. Side conditions (such as banning vehicles from visiting incompatible customers or the vehicle capacity) are ensured within the SAM algorithm. Improvements on the routing costs due to the orientation of the routes are done locally after the assignment of the nodes that a vehicle with visit is done.

At each round a classical capacitated VRP (with unlimited and homogeneous fleet) is solved. The sub-problems are efficiently solved with a randomized version of the Clarke and Wright savings (CWS) heuristic. The CWS heuristic uses the concept of savings which is the driver for merging routes. The savings reflect the gains in terms of visiting two consecutive nodes instead of going back and forth the depot for each of them. For example, if we consider just two nodes, we have two alternatives. One is to serve each node individually which reports a total cost of $2c_{0i} + 2c_{0j}$ (assuming symmetric costs). The other is a route that serves first node i, next goes to node j and returns back to the depot, which costs $c_{0i} + c_{ij} + c_{j0}$. The difference between those two alternatives gives the following savings, $S_{ij} = c_{0i} + c_{0j} - c_{ij}$. These savings are used to select the next arc when building the routes. For a deeper explanation, the reader is referred to [18].

Figure 1 shows the scheme used to solve the heterogeneous, asymmetric and site-dependent VRP. It starts by defining the set of nodes that need to be routed. The saving cost for each link in the network is calculated. At this initial stage, the network is reduced to a non-oriented symmetric network. A weighted saving associated to the link that connects node i and j is computed as defined in [4]:

$$\hat{S}_{ij} = \beta\max\{S_{ij}, S_{ji}\} + (1 - \beta)\min\{S_{ij}, S_{ji}\}, \quad \beta \in [0.5, 1] \tag{1}$$

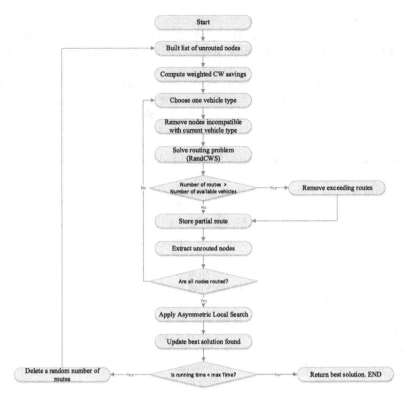

Fig. 1. Flowchart of the SAM for the asymmetric, heterogeneous and site-dependent vehicle routing problem

In Eq. (1), S_{ij} is the saving associated to the arc (i, j) and S_{ji} is the saving associated to the arc (j, i).

Then an available vehicle is chosen and the set of nodes to be routed is modified by excluding those nodes which are incompatible with the current vehicle type. The classical Clarke and Wright Savings algorithm is run enhanced with a randomization process. Instead of using the best possible edge to build a route, all edges are assigned a probability to be selected. Better edges have higher probabilities than the others. A biased distribution such as the geometric distribution is used. Then, several runs of the same procedure come up with different proposals. The best solution is then chosen as a partial solution.

In this partial solution, it is likely that the number of vehicles used is higher than the available ones. In this case, a route for each available vehicle is selected. The other routes are removed from the partial solution. In the next round, those nodes that remain unrouted plus any other node left out because of vehicle incompatibility compound the subset of nodes to be routed.

When all nodes have been assigned to a proper vehicle, a local search procedure is applied to improve the visiting order within each route. At this point asymmetric costs are employed. A first approach is to check a given route in both directions and take that with lowest value. More advanced techniques can be used such as those presented in [4].

The scheme presented so far constitutes a basic local search heuristic that determines a set of routes ensuring that vehicles capacity is satisfied (CWS take care of the capacity constraint), customers are served with adequate vehicles (incompatible nodes are removed from the set of unrouted nodes) and routes are measured with asymmetric costs (using a specific local search). To generate new solutions, a base solution is partially destroyed (by removing a random number of routes) and the procedure is repeated.

5 Numerical Experiments

The aforementioned procedure was coded as a Java application. In order to test the potential of our method we carried out three different experiments: (*i*) homogeneous VRP with asymmetric instances, (*ii*) heterogeneous VRP with asymmetric distances and, (*iii*) heterogeneous VRP with asymmetric distances and site-dependency (*i.e.* some customers can not be served by the vehicles with the largest capacity). Each experiment was carried over all selected instances using 5 different random seeds and was executed using a 2.4 GHz core i5 personal computer with 8 Gb RAM. The execution time was established in 120 s per run.

5.1 Test Cases

In order to perform the tests described above, we have randomly selected 4 instances with Euclidean distances from classical Homogeneous VRP instances available at [19]. These instances have been modified using the following procedure:

1. For each pair of nodes (*a*, *b*), if the *y*-coordinate of *b* is greater than *y*-coordinate of *a* we multiply the distance of the arc by 1.1, otherwise the distance remains unmodified. This allows us to generate asymmetric instances that can be easily replicated in further research projects.
2. Next we generate two new vehicle capacities corresponding to the 75 % and 50 % of the original (T1) vehicle capacity. The number of available vehicles of the original capacity is chosen to be lower than the number of vehicles needed in the best known solution (BKS) for the homogeneous case with asymmetric distances.
3. In order to have site-dependency, for each instance we randomly select a sub-region of the whole x-y space and we restrict the nodes belonging to that area of being visited by T1-capacity vehicles.
4. The number of available vehicles for each of the new capacities is established to ensure the demand satisfaction constraints, considering a reduced fleet of T1-capacity trucks and the demand of the restricted nodes.

Table 1. Summary of results

Instances	Distance based costs (symmetric) BKS	Homogeneous case					Heterogeneous case										Site dependent						
		Avail. trucks	OBS total cost (2)	OBS distance-based cost (3)	Used trucks	GAP % (3)-(1)	Instance information			Non-sit dependent													
							Available trucks / Capacity			OBS total cost (4)	OBS distance-based cost (5)	GAP % (4)-(2)	GAP % (5)-(3)	Used trucks			OBS total cost (6)	OBS distance-based cost (7)	GAP % (6)-(2)	GAP % (7)-(3)	Used trucks		
							T1	T2	T3					T1	T2	T3					T1	T2	T3
P-n40-k5	461.73	8	2115.45	471.82	5	2.18	4 / 140	2 / 105	3 / 70	2045.87	475.62	-3.29	0.81	4	1	0	2275.93	575.89	7.58	22.06	3	2	1
B-n41-k6	834.92	13	3089.90	829.97	6	-0.59	5 / 100	6 / 75	6 / 50	3095.45	975.26	0.17	17.51	3	4	0	3513.71	1072.76	13.72	29.25	4	4	0
B-n45-k5	754.22	8	2729.91	743.30	5	-1.44	4 / 100	3 / 75	3 / 50	2579.33	791.69	-5.51	6.51	3	3	0	2748.16	829.38	0.67	11.58	3	3	0
A-n80-k10	1766.50	16	6334.66	1778.22	10	0.66	8 / 100	6 / 75	6 / 50	6508.37	1969.47	2.74	10.76	8	3	0	6627.54	2062.12	4.62	15.97	6	5	0
AVERAGE GAP						0.20						-1.47	8.89						6.64	19.71			

5.2   Results and Analysis

Table 1 summarizes, for each instance, the results obtained during the experiments. For each instance we report the best solution obtained in terms of total costs (vehicle fixed cost + vehicle variable cost) and its corresponding distance-based cost. Note that our objective function is the minimization of the total costs, but we used the distance-based cost only for comparison purposes with respect to the best known solutions for the symmetric case.

The left side of the table allows the comparison of the asymmetric and symmetric versions of the homogeneous VRP. In this case, since we have only one type of vehicles, the optimization of the total costs has the same solution than the optimization of routing distances. Our mean gap in terms of distance cost is 0.20 % which shows the competitiveness of our approach.

The right side of the table shows the results obtained for the heterogeneous VRP without and with site-dependency, which corresponds to experiments (*ii*) and (*iii*) respectively. In the case of non-site dependency, we can see that, in average, the heterogeneous version outperforms the homogeneous version in terms of total costs (average gap of **−1.47 %**) while the distance cost is **8.89 %** higher, in average. Since there are less available vehicles with larger capacities, solutions are forced to use smaller vehicles (with lower fixed and variable costs) and to perform more trips. In the case of site-dependency, which restricts even more the usage of larger trucks, the total costs increases (average gap of **6.64 %**) with respect to the homogeneous case, while the distance costs increase, on average **19.71 %**.

6   Conclusions and Future Enhancements

We have introduced an efficient, fast and easy to implement multi-round algorithm for planning goods delivery in mountainous regions with heterogeneous fleet. This situation is represented in this article by the so-called Heterogeneous Site-Dependent with Asymmetric Costs Vehicle Routing Problem (HSDA-VRP). The proposed algorithm is based on a randomized version of the CWS heuristic which assigns a higher probability of being chosen to the most promising movements. Preliminary tests carried out show that our approach seems promising in order to solve more realistic versions of the VRP.

Further research efforts could be oriented to include real-life data or conditions (i.e. customer locations, real distance-based costs, other vehicle types and associated costs, uncertain demand, uncertain travel times, etc.)

Acknowledgements.   This work has been partially supported by the Spanish Ministry of Economy and Competitiveness (TRA2013-48180-C3-P and TRA2015-71883-REDT), FEDER, the Catalan Government (2014-CTP-00001) and the Government of Andorra (ACTP022-AND/2014). Likewise, we want to acknowledge the support received by the Special Patrimonial Fund from Universidad de La Sabana and the doctoral grant from the UOC.

References

1. Juan, A., Faulin, J., Cruz, J.C., Barrios, B.B., Martinez, E.: A successive approximations method for the heterogeneous vehicle routing problem: analysing different fleet configurations. Eur. J. Ind. Eng. **8**(6), 762 (2014)
2. Juan, A., Goentzel, J., Bektaş, T.: Routing fleets with multiple driving ranges: is it possible to use greener fleet configurations? Appl. Soft Comput. **21**, 84–94 (2014)
3. Buhrkal, K., Larsen, A., Ropke, S.: The waste collection vehicle routing problem with time windows in a city logistics context. Procedia Soc. Behav. Sci. **39**, 241–254 (2012)
4. Herrero, R., Rodríguez, A., Cáceres-Cruz, J., Juan, A.A.: Solving vehicle routing problems with asymmetric costs and heterogeneous fleets. Int. J. Adv. Oper. Manag. **6**(1), 58–80 (2014)
5. Bodin, L.D.: Twenty years of routing and scheduling. Oper. Res. **38**(4), 571–579 (1990)
6. Laporte, G., Osman, I.H.: Routing problems: a bibliography. Ann. Oper. Res. **61**(1), 227–262 (1995)
7. Toth, P., Vigo, D.: An overview of vehicle routing problems. Discret. Appl. Math. **123**(1–3), 1–26 (2002)
8. Golden, B., Raghavan, S., Wasil, E.: The Vehicle Routing Problem: Latest Advances and New Challenges, vol. 43 (2008)
9. Hoff, A., Andersson, H., Christiansen, M., Hasle, G., Løkketangen, A.: Industrial aspects and literature survey: combined inventory management and routing. Comput. Oper. Res. **37**(9), 1515–1536 (2010)
10. Lin, C., Choy, K.L., Ho, G.T.S., Chung, S.H., Lam, H.Y.: Survey of green vehicle routing problem: past and future trends. Expert Syst. Appl. **41**(4), 1118–1138 (2014)
11. Kallehauge, B.: Formulations and exact algorithms for the vehicle routing problem with time windows. Comput. Oper. Res. **35**(7), 2307–2330 (2008)
12. Baldacci, R., Mingozzi, A., Roberti, R.: Recent exact algorithms for solving the vehicle routing problem under capacity and time window constraints. Eur. J. Oper. Res. **218**(1), 1–6 (2012)
13. Cordeau, J.-F., Laporte, G.: A tabu search algorithm for the site dependent vehicle routing problem with time windows. INFOR **39**(3), 292–298 (2001)
14. Nag, B., Golden, B., Assad, A.: Vehicle routing with site dependencies. In: Golden, B., Assad, A. (eds.) Vehicle Routing: Methods and Studies, pp. 149–159. Elsevier, Amsterdam (1988)
15. Chao, I.M., Liou, T.S.: A New Tabu Search Heuristic for the Site-Dependent Vehicle Routing Problem. Next Wave Comput. Optim. Decis. Technol. **29**, 107–119 (2005)
16. Pisinger, D., Ropke, S.: A general heuristic for vehicle routing problems. Comput. Oper. Res. **34**(8), 2403–2435 (2007)
17. Yusuf, I.: Solving multi-depot, heterogeneous, site dependent and asymmetric VRP using three steps heuristic. J. Algorithms Optim. **2**(2), 28–42 (2014)
18. Juan, A.A., Faulin, J., Ruiz, R., Barrios, B., Caballé, S.: The SR-GCWS hybrid algorithm for solving the capacitated vehicle routing problem. Appl. Soft Comput. **10**(1), 215–224 (2010)
19. BranchandCut VRP Data. http://www.coin-or.org/SYMPHONY/branchandcut/VRP/data/. Accessed 25 Jan 2016

Checking the Robustness of a Responsible Government Index: Uncertainty and Sensitivity Analysis

Manuel Urquijo-Urrutia[1], Juan Diego Paredes-Gázquez[2(✉)],
and José Miguel Rodríguez-Fernández[3]

[1] Universidad del País Vasco Vasco-Euskal Herriko Unibertsitatea, Leioa, Spain
M-Urquijo@euskadi.eus
[2] Universidad Nacional de Educación a Distancia, Madrid, Spain
juandiegoparedes@cee.uned.es
[3] Universidad de Valladolid, Valladolid, Spain
jmrodrig@eco.uva.es

Abstract. The Basque Country has many public entities ruled by private law, but there is no measure about how they conduct a responsible corporate governance. We develop a Responsible Government Index (RGI) as a tool for (1) assessing and managing the global situation regarding public entities in the Basque Country and (2) making the necessary decisions to attend the stakeholders of these entities. The evidence from this study implies that the RGI is a robust composite measure.

Keywords: Composite index · Uncertainty analysis · Sensitivity analysis · Variance

1 Introduction

The government of the Basque Country, an Autonomous Community in northern Spain, holds many publicly owned entities ruled by private law, such as public companies, public foundations and consortiums. The public contribution has been decisive in the constitution and the daily operations of these entities. Therefore, it is important to know if they conduct a responsible corporate governance, which is a combination of good governance and social responsibility.

We develop a Responsible Government Index (RGI) as a tool for (1) assessing and managing the global situation regarding public entities in the Basque Country and (2) making the necessary decisions to attend the stakeholders of these entities.

In order to develop the RGI, we followed a composite index (CI) construction method similar to those proposed in the Handbook on Constructing Composite Indicators [1]. Other studies also follow this method for constructing CIs [2, 3]. One of the steps of these guidelines checks the robustness of CIs through uncertainty and sensitivity analyses, which are methods based on sampling and simulations. We explain how to conduct these analyses, showing that they are essential steps in the construction of a CI.

© Springer International Publishing Switzerland 2016
R. León et al. (Eds.): MS 2016, LNBIP 254, pp. 133–142, 2016.
DOI: 10.1007/978-3-319-40506-3_14

2 Theoretical Background

Corporate Social Responsibility and Corporate governance are two different but related concepts. On the one hand, the firms, as the citizens, have rights and responsibilities. Corporate Social Responsibility (CSR) are the duties of the firm towards the society. On the other hand, corporate governance is the system of direction, control and evaluation of firms.

The combination of CSR and corporate governance results in the Responsible Government concept, which is a balanced mix of policies in those areas meeting the demands and expectations of stakeholders. The Public Administrations should consider de demands and expectations of stakeholders affected by their decisions [4, 5]. In this sense, Governments often consult with their stakeholders the laws they issue. At the same time, these Governments should balance the participation and influence of their stakeholders, avoiding the concentration of power in one or some stakeholders. The Public Administrations should ensure fair rules for dealing with stakeholders.

The Government of Spain and some Autonomous Communities have issued laws and recommendations concerning codes of ethics and conduct, transparency, conflicts of interest, social responsibility and good governance of senior officials of the Public Administration. However, there is no reference in codes covering both corporate governance and CSR of publicly owned entities. The approval by the Basque Country Government of a Guide to Responsible Government constitutes an unprecedented public sector in Spain [6].

The publicly owned entities of the Basque Country fulfill many of the OECD corporate governance recommendations for publicly owned entities [7]. It is therefore important to know up to what point these entities implement a Responsible Government. Thus, we develop a Responsible Government Index (RGI) as a tool for (1) assessing and managing the global situation regarding public entities in the Basque Country and (2) making the necessary decisions to attend the stakeholders of these entities.

3 Responsible Government Index Construction

3.1 Data

The empirical study includes 26 entities from public sector, covering a large percentage of the value of the economic activities of the Basque Country. The data has been obtained by filling in questionnaires, covering a period of ten quarters between 2009 and 2012. There are no missing data. The questionnaires collect information about eleven variables or indicators. Table 1 reports the descriptive statistics of the variables.

3.2 Multivariate Analyses

The dual version of the STATIS method (STATIS means Structuration des Tableaux á Trois Indices de la Statistique) examines the structure of the data and condense the information of the ten periods they provide [8, 9]. A rotated principal component

Table 1. Descriptive statistics

Variable	Min.	Max.	Average	Std. dev.
Transparency	0	6.571	3.198	1.662
Board of directors	1.300	6.400	3.483	1.511
Directors	0.769	8.385	4.234	2.424
Ethics code	0	7.714	2.148	2.420
CSR strategy	0	10	2.185	2.457
Clients	0	7.931	5.231	2.237
Suppliers	0	8.889	3.923	2.695
Energy and environment	0	8.750	3.875	2.761
Employees	2.952	9.524	6.068	1.780
Society	0	8.571	3.236	2.543
Engagement with stakeholders	0	10	3.679	3.079

analysis eases the interpretation of the STATIS dual results. Table 2 shows the eigenvalues obtained. The adequacy measures indicate that the method summarizes the data properly (KMO = 0.731; Bartlett $\chi^2_{55} = 1648.977$, p < 0.000).

Table 2. Eigenvalues after Varimax rotation

Factor	Eigen.	% variance	% cumulative	Factor	Eigen.	% variance	% cumulative
F1	4.47	40.67	40.67	F6	0.54	4.99	89.34
F2	2.27	20.71	61.38	F7	0.35	3.19	92.53
F3	1.12	10.18	71.56	F8	0.28	2.57	95.11
F4	0.81	7.38	78.95	F9	0.24	2.26	97.37
F5	0.59	5.40	84.35	F10	0.16	1.47	98.85
				F11	0.12	1.14	100

In order to construct the RGI, we select the factors fulfilling three conditions [10, 11]:

- Eigenvalues greater than one.
- Percentage of variance explained greater than 10 %.
- Percentage of cumulative variance explained greater than 60 %.

The first four factors fulfill these conditions. Table 3 details the factor loadings of the first four factors.

After the Varimax rotation, we interpret the four factors or dimensions. D1 factor represents the Stakeholders dimension (variables with highest factor loadings: Clients, Suppliers, Energy and environment, Employees and Society). D2 factor contains Management issues (variables with highest factor loadings: Board of Directors and Directors). D3 is the Ethics dimension (variable with highest factor loading: Ethics Code). Finally, D4 represents the CSR Strategy dimension (variables with highest factor loadings: Transparency, CSR Strategy and Engagement with stakeholders).

Table 3. Factor loadings after Varimax rotation

Variable	D1	D2	D3	D4
Transparency	0.193	0.419	0.178	0.676
Board of directors	0.045	0.856	0.078	0.274
Directors	−0.126	0.899	0.220	0.007
Ethics code	−0.011	0.355	0.853	0.164
CSR strategy	0.300	−0.016	0.400	0.745
Clients	0.752	0.457	−0.193	0.092
Suppliers	0.748	−0.055	0.493	0.111
Energy and environment	0.811	−0.139	0.148	0.267
Employees	0.695	−0.178	0.175	0.488
Society	0.792	0.037	−0.192	0.272
Engagement with stakeholders	0.260	0.177	−0.096	0.756

The weights of the dimensions are obtained by following this process [10]:

1. Square the factor loadings of the variables with greater factor loadings. Table 4 shows the squared factor loading of factor loadings greater than 0.6.
2. Calculate the relative contribution of the variables to each factor, which sets the weight of the variables. The relative contribution is the square factor loading of the variable divided by the sum of squared factor loadings of the factor. Table 4 repots the relative contribution of the variables. The score of each dimension is calculated as Eq. 1 shows, where e denotes the entity and t the quarter or period.

$$
\begin{aligned}
D1(e, t) &= 0, 20 * CL(e, t) + 0, 19 * SP(e, t) + 0, 22 * EE(e, t) \\
&\quad + 0, 17 * EM(e, t) + 0, 22 * SO(e, t) \\
D2(e, t) &= 0, 48 * BD(e, t) + 0, 52 * D(e, t) \\
D3(e, t) &= EC(e, t) \\
D4(e, t) &= 0, 29 * TR(e, t) + 0, 35 * CS(e, t) + 0, 36 * ES(e, t)
\end{aligned}
\tag{1}
$$

3. Calculate the relative dimension contribution of each factor to the total variability of the factors, which sets the weight of the factors. The relative dimension contribution is the sum of squared factor loadings of the factor divided by the sum of squared factor loadings of all the factors. Table 4 shows the relative dimension contribution of the factors, which constitutes the weight of each dimension in the final RGI. The final weights of the dimensions are: Stakeholders (D1 = 42,9 %), Management (D2 = 22,8 %), Ethics (D3 = 10,8 %) and Transparent strategy (D4 = 23,5 %).
4. Calculate the RGI. The final scores of the RGI combines the two previous steps in Eq. 2.

$$
\begin{aligned}
RGI(e, t) &= 0, 429 * D1(e, t) + 0, 228 * D2(e, t) + 0, 108 * D3(e, t) \\
&\quad + 0, 235 * D4(e, t)
\end{aligned}
\tag{2}
$$

Note that we have constructed both a CI for each dimension and a global RGI. The rank of entities is not reported due to confidentiality issues.

Table 4. Squared factor loadings after Varimax rotation

Variable	D1		D2		D3		D4	
	Load.	Cont.	Load.	Cont.	Load.	Cont.	Load.	Cont.
Transparency (TR)							0.45	29 %
Board of directors (BD)			0.73	48 %				
Directors (DI)			0.80	52 %				
Ethics code (EC)					0.72	100 %		
CSR strategy (CS)							0.55	35 %
Clients (CL)	0.56	20 %						
Suppliers (SP)	0.55	19 %						
Energy and environment (EE)	0.65	22 %						
Employees (EM)	0.48	17 %						
Society (SO)	0.62	22 %						
Engagement with stakeholders (ES)							0.57	36 %
Sum	2.89		1.54		0.72		1.58	
Dimension contribution	0.429		0.228		0.108		0.235	

Load.: square factor loading
Cont.: relative contribution

4 Robustness Check

In order to assess if the RGI is robust, we conduct uncertainty (UA) and sensitivity analyses (SA) to the RGI scores in period ten. These analysis asses the change in the CI score attributed to subjective choices made during its construction (robustness). UA focuses on how uncertainty in the input factors propagates through the structure of the CI and affects its values. SA studies how much each individual source of uncertainty contributes to the output variance [12].

These two analyses require defining input factors, which are hypothetic combinations of the subjective choices adopted across the construction process of the RGI. Among the frequent input factors are normalization, weighting method, aggregation procedure and missing data treatment. The next circumstances limited the subjective choices we made:

- Indicators are already normalized, since their values comes from a rotated principal component analysis.
- The geometric aggregation procedure is not appropriate. Those entities with a score of zero in one or more dimensions, automatically get a score of zero on the CI.
- There are no missing data.

Thus, UA and SA focus on the weighting of the dimensions of the CI.

4.1   Uncertainty Analysis

We define four input factors, one for each dimension d. The values for each dimension are weighted according to a trigger X_D that is sampled from a uniform distribution (U), where X is the random variable associated to the dimension D. The interval of the distribution, which is different for each dimension, includes any value 50 % above and below the initial weight of the dimension (Table 5).

Table 5. Input factors

Dimension	Initial weight	Input factor distribution
D1: Stakeholders	0,429	$X_{D1} \sim U(0,210–0,630)$
D2: Management	0,228	$X_{D2} \sim U(0,114–0,342)$
D3: Ethics	0,108	$X_{D3} \sim U(0,054–0,162)$
D4: Transparent strategy	0,235	$X_{D4} \sim U(0,117–0,352)$

A quasi-random sampling procedure generates the combinations of input factors [13–15]. The sample size is $n = 124$ for each input factor, so we get a total of 2n (k +1) = 1240 simulations, where $k = 4$ is the number of input factors. Sampling is done using SimLab [16]. The results of the UA are obtained as follows:

1. Compute a new score of the CI for each entity in each simulation.
2. Calculate the new rank of the entities in each simulation.
3. Calculate the difference between its original position in the rank (the one we got previously to the UA) and the new position in the simulation for each entity. The desirable value of this difference is zero; it means that the rank of the entity is not affected by the weighs of the dimensions of the CI.

The Fig. 1 summarizes the position shifts of all the entities. Since there are 1240 simulations and 26 entities, the total number of position shifts is 1240 * 26 = 32240. The most frequent value is zero, so there is no position shift in most of the cases. A shift of four or more positions is rare.

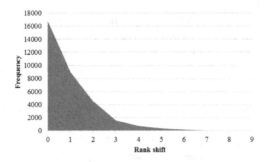

Fig. 1. Frequency of rank shifts

The Fig. 2 represents boxplots, which visually summarizes the rank shifts of each entity across the 1240 simulations. The range of the boxplot indicates the size of the shift: the greater the size, the more sensible the entity is to a different weighting scheme. The boxplots show that:

- There are three entities (the first and the last two) whose rank does not change in the simulations. In these cases, the CI is fully robust.
- The median is zero for 22 entities: their rank is the same in at least 512 simulations.
- The median is one for four entities: they change just one position in at least 512 simulations.
- There are entities whose rank changes five positions or more, regardless of the sign of the shift. However, as observed in Fig. 1, shifts greater than three or more positions are infrequent compared to the total number of shifts.

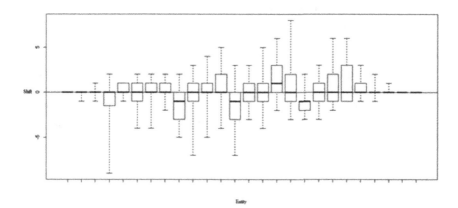

Fig. 2. Uncertainty analysis

4.2 Sensitivity Analysis

SA decomposes the variance of the variable Y, which in our case is the RGI, following Eq. 3. S_i is the contribution of each input factor to the change in the score of the RGI or first-order effect. $V(Y)$ is the total effect, which includes the conditional variances V_{ij} corresponding to more than one input factor.

$$S_i = V_i/V = V[E[Y|X_i]]/V[Y]$$
$$V_i = V_{X_i}\{E_{X_{\_i}}(Y|X_i)\}$$
$$V_{ij} = V_{X_iX_j}\{E_{X_{\_ij}}(Y|X_i, X_j)\} - V_{X_i}\{E_{X_{\_j}}(Y|X_i)\} - V_{X_i}\{E_{X_{\_j}}(Y|X_j)\} \quad (3)$$
$$V(Y) = \sum_i V_i + \sum_i \sum_{j > i} V_{ij} + \ldots + V_{12\ldots k}$$

The SA is estimated using the updated algorithm of Sobol in SimLab [14, 15]. The input variables are the four input factors, while the output variable R_s is the average of the absolute differences in entities' ranks with respect to a reference ranking over the $M = 26$ entities (Eq. 4). The reference ranking is the original position of the entity in the RGI.

$$R_s = (\sum\nolimits_{c=1}^{M} |\text{rank}_{\text{ref}}(Y_c) - \text{rank}(Y_c)|)/M \tag{4}$$

Table 6 presents the first-order effects and the total effects of the SA. There is no rule establishing to what extent an effect is important. An input factor important if it explains more than $1/k$ of the output variance [12].

Table 6. Sensitivity measures

Input factor	Dimension	First-order effect (S_i)	Total effect (S_{Ti})	S_{Ti} - S_i
X_{D1}	D1: Stakeholders	0,222	0,732	0,510
X_{D2}	D2: Management	0,150	0,589	0,438
X_{D3}	D3: Ethics	0,066	0,157	0,091
X_{D4}	D4: T. Strategy	0,210	0,236	0,026
	Sum	0,649	1,714	

The input factor that most affects the average shift in the rank of the entities corresponds to the Stakeholders dimension, and the one that least affects corresponds to the Ethics dimension. The values of the first-order effects are below the $1/k$ threshold. The difference between the total effects and first-order effects, which captures the conditional variances corresponding to more than one factor, causes most of the variability of the average shift.

5 Discussion

In recent years, there has been considerable interest in CIs [17]. These tools help in the policymaking and evaluation process. However, CIs must not be constructed any which way. A non-robust CI may lead to wrong decisions. The construction process could follow a method similar to those proposed in the Handbook on Constructing Composite Indicators.

Although some authors have suggested using uncertainty and sensitivity analyses to test the robustness of CIs, such analysis are seldom carried out [12, 18]. We check the robustness of the RGI using both uncertainty and sensitivity analyses. We observe that:

- Frequently, there is no shift in the rank of the entities, or they change just one or two positions (Fig. 1). Moreover, as the boxplot shows (Fig. 2), almost all the medians are zero, so the rank does not usually change if the weighting of the CI is different.
- The values of the firs-order effects are below the $1/k$ threshold, so there is no threat to the robustness of the CI (Table 6). The simulations reveal no input factors significantly affecting the variation of CI.
- The rank shifts obey to difference between the total effects and first-order effects (Table 6).

Therefore, based on the input factors defined for the analyses, we conclude that the RGI is robust.

The RGI is and index whose first virtue is its intuitive simplicity, allowing and effective communication with stakeholders from diverse backgrounds. The RGI is also a tool that provides a clear and unequivocal message in a consolidated single score in a given time for each entity, facilitating comparison with other entities.

6 Conclusion

Publicly owned entities play an important role in the Basque Country economy. In order to measure how responsible is the management of these entities, we have constructed a composite index: the RGI. To enable us to check the robustness of the RGI, we performed uncertainty and sensitivity analyses. These methods are based on sampling of input factors and simulation of their combinations. The evidence from this study implies that, for the input factors we defined, the RGI is a robust composite measure.

The main limitation of the study is the small sample size. Additionally, the weighting of indicators is valid just for the ten quarters period considered, so if the data are updated with new years it is required a new estimation of the weights of the indicators. Concerning future lines of research, a point of interest is conducting longitudinal studies to analyze the impact of the practices of responsible government in the management of the entities. The interaction between the weaknesses in the dimensions of the RGI and the measures adopted by entities to improve their score is very interesting. These relationships of cause and effect can be observed in the future through the continued publication of the scores of the RGI and the periodic sending of questionnaires to entities to meet their performance.

References

1. OECD: Handbook on constructing composite indicators. Methodology and userguide. OECD Publications, Paris (2008)
2. Gaeta, G.L., Ercolano, S.: Reassesing the power of purse. A new metodology for the analysis of the institutional capacity for legislative control over the budget. J. Dyses **3**, 54–75 (2010)
3. Romano, O., Ercolano, S.: Who Makes the Most? Measuring the Urban Environmental Virtuosity. Springer Science+Business (2012)
4. Melle, M.: La Responsabilidad Social dentro del Sector público. Revista Ekonomiaz **65**, 85–107 (2007)
5. Friedman, A.L., Miles, S.: Stakeholders Theory and Practice. Oxford University Press, Oxford (2006)
6. Basque Country Autonomous Community: Acuerdo de Consejo de Gobierno Vasco de 21 de abril de 2009
7. OCDE: Principles of Corporate Governance. OECD Publications, Paris (2004)

8. L'Hermier des Plantes, H.: Structuration des Tableaux á Trois Indices de la Statistique. Tesis. Université de Montpellier (1976)

9. Lavit, C.: Application de la Méthode STATIS. Stat Anal Donnés. **10**, 103–116 (1985)

10. Nicoletti, G., Scarpetta, S., Boylaud, O.: Summary Indicators of Product Market Regulation with an Extension to Employment Protection Legislation. In: OCDE, Economics Department, Working Papers no. 226 (2000)

11. Mascherini, M.: Composite indicators. development and assessment. In: EUSTAT International Statistics Seminar, Bilbao (2009)

12. Saisana, M., Saltelli, A., Tarantola, S.: Uncertainty and sensitivity analysis techniques as tools for the quality assessment of composite indicators. J. Roy. Stat. Soc.: Ser. A **168**, 307–323 (2005)

13. Sobol, I.M.: On the distribution of points in a cube and the approximate evaluation of integrals. USSR Comput. Math. Phys. **7**, 86–112 (1967)

14. Sobol, I.M.: Sensitivity analysis for non-linear mathematical models. Math. Mod. Comput. Exp. **1**, 407–414 (1993)

15. Saltelli, A.: Making best use of model valuations to compute sensitivity indices. Comput. Phys. Communs **145**, 280–297 (2002)

16. Saltelli, A., Tarantola, S., Campolongo, F., Ratto, M.: Sensitivity Analysis in Practice, a Guide to Assessing Scientific Models. Wiley, New York (2004)

17. Paruolo, P., Saisana, M., Saltelli, A.: Ratings and rankings: Voodoo or science? J. Roy. Stat. Soc.: Ser. A **176**, 609–634 (2013)

18. Singh, R.K., Murty, H.R., Gupta, S.K., Dikshit, A.K.: An overview of sustainability assessment methodologies. Ecol. Ind. **15**, 281–299 (2012)

Engineering and Other General Applications

Using Biased Randomization for Trajectory Optimization in Robotic Manipulators

Alba Agustín[1($\boxtimes$)], Alberto Olivares[2], and Ernesto Staffetti[2]

[1] Department of Statistics and OR, Public University of Navarre,
Los Magnolios Blg. Campus Arrosadia, 31006 Pamplona, Spain
`albamaria.agustin@unavarra.es`
[2] Department of Information and Communication Technology,
Rey Juan Carlos University, Camino del Molino, s/n, 28943 Fuenlabrada, Spain
`{alberto.olivares,ernesto.staffetti}@urjc.es`

Abstract. We study the problem of optimization of trajectories for a robotic manipulator, with two degrees of freedom, which is constrained to pass through a set of waypoints in the workspace. The aim is to determine the optimal sequence of points and continuous optimal system trajectory. The actual formulation involves an optimal control problem of a dynamic system within integer variables that model the waypoints constrains. The nature of this problem, highly nonlinear and combinatorial, makes it particularly difficult to solve. The proposed method combines a meta-heuristic algorithm to determine the promising sequence of discrete points with a collocation technique to optimize the continuous path of the system. This method does not guarantee the global optimum, but can solve instances of dozens of points in reasonable computation time.

Keywords: Robotics · Optimal control · Motion planning · Meta-heuristics · Biased-randomization

1 Introduction

A fundamental problem in robotics is the task planning problem. Given the models of the robot manipulator and the environment in which it operates, the problem is to generate a sequence of actions to accomplish a given task. For assembly, material handling, spot welding, measuring, testing, and inspecting one wants to generate a continuous intersection-free motion of the robot manipulator that connects several given configurations of the end-effector.

In this work, the energy-optimal motion planning problem for planar robot manipulators with two revolute joints is studied. In addition, the end-effector of the robot manipulator is constrained to pass through a set of waypoints, whose sequence is not predefined. We propose a multi-start approach to solve the problem that determine the promising discrete path and evaluate the continuous dynamic trajectory.

The paper is organized as follows. The robot motion planning problem studied in this paper will be stated in Sect. 2. Our Multi-Start approach is described in Sect. 3,

© Springer International Publishing Switzerland 2016
R. León et al. (Eds.): MS 2016, LNBIP 254, pp. 145–154, 2016.
DOI: 10.1007/978-3-319-40506-3_15

for which we previously motivate the basis. In Sect. 4 the results obtained applying our approach to several instances are reported. Finally, some conclusions will be drawn in Sect. 5.

2   The Robot Motion Planning Problem

The motion planning problem is an optimal control problem of a mechanical system. Each system comprises a dynamic model to take into account [1, 2]. That is, besides geometrical feasibility, it is also important to ensure dynamical feasibility. Related to optimality, the motion planning must be executed with minimum energy consumption. Furthermore, being a dynamic system, the solution of the problem must provide the optimal scheme of accelerations and velocities during the motion.

In particular, the mechanical system for which we attempt optimal control is a planar robot manipulator with two revolute joints, which we will denote as RR. The RR robot qualitatively corresponds to the model of the first two links of a SCARA (Selective Compliant Assembly Robot Arm) without taking into account the vertical one, see the left-hand side of Fig. 1. And, however this simple robot manipulator can be, it has a very complex nonlinear dynamics and comprises most of the kinematical and dynamical properties of a typical industrial robot. So, the RR is composed by two homogeneous links and two actuated joints moving in a horizontal plane $\{x, y\}$, as shown in the right-hand side of Fig. 1, where l_i is the length of link i, r_i is the distance between joint i and the mass center of link i, and θ_i is the angular position of link i, for $i = 1, 2$. Finally, the vector $\tau = (\tau_1, \tau_2)$ defines the control inputs of the system, where τ_1 is the torque applied by the actuator at joint 1 and τ_2 is the torque applied by the actuator at joint 2.

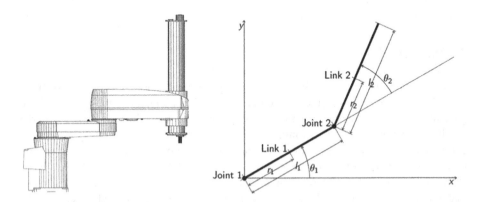

Fig. 1. A robot manipulator that moves in a horizontal plane

In this work, the RR motion planning problem not only includes its correspondent dynamic model, but also constraints between the initial and final positions. In partic- ular, the robot manipulator is constrained to pass once through all the given points in the workspace, with no predefined sequence, when moving from an initial position p_I to a final position p_F. The presence of these waypoint constraints adds a combinatorial complexity to this optimal control problem and makes it particularly difficult to solve. In summary, the main features that make this problem especially complex: (i) dynamic problem (ii) non-linear problem and (iii) combinatorial problem.

3 How to Solve the Problem

We use this section to highlight what it is expected to be a solution of the problem. According to that, we analyze the problem complexities and motivate our approach based on split and conquer strategy. In particular, we split as follows: first, elaborate a Multi-Start algorithm for the combinatorial problem and then, display the nonlinear problem formulation in the IPOPT solver to obtain the dynamic motion of the problem.

3.1 Background and Motivation

In Bonami *et al.* [3] they model the problem as a Mix Integer NonLinear Programming. Moreover, the problem is converted into a NonLinear Programming (NLP) problem, when the sequence of waypoints is fixed. So, Bonami *et al.* [3] solve the *RR* optimal control problem for 18 waypoints (see Fig. 2) using BONMIN solver [4], which integrates a BB algorithm and the IPOPT solver [5] for NLP. It is important to point out that, since both the order in which the waypoints are visited and the corresponding velocities are not specified, they must be determined. As far as we know, for the latter it is necessary the use of optimization engines such as IPOPT to solve the NLP.

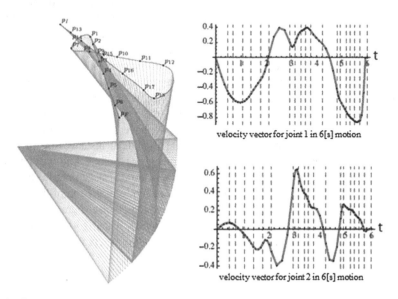

Fig. 2. BONMIN pseudo-optimal trajectory and control variables within 18 waypoints.

On the other hand, we observe that the continuous path is strongly affected by the discrete path, i.e. the sequence of points. In particular, we experimentally checked that main energy consumption is due to changes in velocity which occur when there is a change in the direction of the motion. Therefore, the preferable sequence of waypoints for minimizing energy consumption will comprise straight paths within smooth turns.

So, in our approach will take advantage of that property and look for promising discrete paths, unlike the BB algorithm [3] which does not make distinction between them. In other words, our approach is based on finding these straight paths and avoiding sharp turns, taking into account the location of the waypoints. Once the promising discrete path have been found, the corresponding continuous optimal trajectories (i.e. continuous paths plus velocity profile along them) is found by solving the NLP problem. In fact, when the sequence of waypoints is fixed, IPOPT solver finds for the NLP problem good dynamic trajectories in short computational times.

3.2   Multi-Start Approach

In order to construct sequences of waypoints which include the minimum changes of direction as possible, we will look for straight segments and then join the segments and the isolated points. The pseudocode for this algorithm is depicted in Fig. 3, and basically, this approach includes the procedures that follow:

1. *Search possible segments in the workspace.* In this procedure we consider each edge from the workspace and compute one by one the acute angle respect to all the other waypoints. In case we obtain 180° the waypoint will be included. We find a segment when there are at least two more waypoints included in the edge.
2. *Use biased randomization.* Once we obtain the list of segments, we sort that list using a skewed probability distribution. A skewed distribution is used here in order to assign higher probabilities of being in the top of the list by segments composed with a larger number of waypoints. In our case, a Geometric distribution with $\alpha = 0.25$ was employed to induce this biased-randomization behavior [6]. So, we always select the segment in the top to be the discrete path solution. Next, we update the list of segments, since the waypoints from the segment selected must be removed from the other segments. Then, sort the list and pick the one in the top again. Step 2 ends when the list of segments is empty.
3. *Join segments and isolated waypoints.* Once we have a "good" selection of segments to be in the discrete path, we compute Euclidean distance for all possible connections, i.e. ends of the segments and isolated waypoints. And we sort the list of possible connections using a biased-randomization of a geometric distribution with $\beta = 0.25$. So, we always join using the connection in the top of the Euclidean distance list. Next, we update the list of connections, since there would be connections that must be removed after we have joined. Then, sort the list and join using the one in the top again. Step 3 ends when all the waypoints are connected in a discrete path.

4. *Repeat step 2 and 3.* After having studied the parameter analysis, the proper stopping criteria for this Multi-Start approach for *RR* problems with 12−18 waypoints, are maximum 10−15 iterations for step 2 and maximum 30 iterations for step 3. It means, for each group of selected lines in step 2, we will try 30 different connections in step 3. So, the Multi-Start approach will create 10−15 X 30 discrete paths, although in many cases we obtain a repeated one from previous iterations.

5. *Sort discrete paths.* We calculate the total acute angle in every distinguished discrete path. The total acute angle is the addition of each acute angle, i.e. acute angle of each 3 consecutive waypoints in the discrete path. And, we sort the discrete path from the lowest to the highest total angle computed.

6. *Solution of the RR.* We call IPOPT through the correspondent .nl file within the NLP problem formulation. Because the discrete path will be fixed according to the list of path, the IPOPT solver provides trajectories quickly. In *RR* problems with 12−18 waypoints, it will be enough checking the first 30 paths from the list.

```
Multi-StartApproach (waypoints, alpha, beta, maxIter2, maxIter3, maxIpopt, nlfile)

% Searching for good discrete paths
edges <- read(waypoints)
segments <- findSegments (edges, waypoints) % step 1
for { 0 to maxIter2 } do
  while { segments list is not empty } do
    randomSelection (segments, alfa) % step 2
    update(segments)
  end while
  distanceList <- calcDist(waypoints isolated & end segments)
  for { 0 to maxIter3 } do
    while { all waypoints not in discrete path} do
      randomSelection(distanceList, beta) % step 3
      update (distanceList)
    end while
    pathList <- save (currentPath) % sorted list of new solutions
  end for % step 4
end for % step 4

% Evaluating the obtained discrete path
calcTotalAngle(pathList)
sortDiscretePaths(pathList) % step 5
energy(bestTrajectory) <- infinite
for { it = 0 to maxIpopt } do
  Trajectory [it] <- solutionIpopt(pathList[it], nlfile) % step 6
  if {energy(Trajectory [it]) < energy(bestTrajectory)} then
    bestTrajectory <- Trajectoty[it]
  end if
end for
return bestTrajectory
```

Fig. 3. Pseudocode for the *RR* pseudo-optimal solution, i.e. continuous trajectory

Finally, to mention that in step 1 the allowable acute angle could be variable. In this case we have considered 180°, i.e. completely straight segments, since the waypoints in the workspace would be somehow aligned. However, in case that the waypoints are not absolutely aligned we just need smoothly decrease the acute angle in order to enable quasi-straight segments.

3.3   Problem Formulation for Ipopt Solver

In this section we summarize the problem formulation of the optimal control problem and the fundamental characteristics studied in [3]. First of all, we present the continuous optimal control problem formulation:

$$\text{Min } J[x(t), u(t), s] := E[x(t_F), s] + \int L[x(t), u(t), s] \, dt \tag{1}$$

subject to:

$$\dot{x}(t) = f[x(t), u(t), s], \quad t \in [t_1, t_F] \tag{2}$$

$$0 = g[x(t), u(t), s], \quad t \in [t_1, t_F] \tag{3}$$

$$0 \le c[x(t), u(t), s], \quad t \in [t_1, t_F] \tag{4}$$

$$r^{ineq}[x(t1), x(t2), \ldots, x(t_{nrineq}), s] \le 0 \tag{5}$$

$$r^{eq}[x(t1), x(t2), \ldots, x(tnreq), s] = 0 \tag{6}$$

$$x(t_I) = x_I \tag{7}$$

$$\psi[x(t_F)] = 0 \tag{8}$$

The objective functional in (1) is given in Bolza form and it is expressed as the sum of the Mayer term, which is assumed to be twice differentiable, and the Lagrange term. Variable $t \in [t_I, t_F]$ represents time, where t_I and t_F are the initial and final time, respectively. $x(t)$ represents the state variables within both, differential and algebraic variables and $u(t)$ represents the control functions, also referred to as control inputs, which are assumed to be measurable. The vector s contains all the time-independent variables of the problem. Equations (2) and (3) represents a Differential Algebraic Equations (DAE) system. The function f is assumed to be piecewise Lipschitz continuous to ensure existence and uniqueness of a solution. The system must satisfy the algebraic path constraints c in (4) and the interior point inequality and equality constraints r^{ineq} and r^{eq} in constraints (5) and (6), respectively, which are assumed to be twice differentiable. Finally, x_I in (7) represents the vector of initial conditions given at the initial time t_I and the function ψ in (8) provides the terminal conditions at the final time t_F, which is assumed to be twice differentiable.

In addition, we can introduce integer variables to obtain a multi-phase problem, and therefore, the so-called mixed integer optimal control problem appears. Moreover, we can convert the mixed-integer optimal control problem into a mixed integer nonlinear programming problem including these transformations: (i) making unknown passage times through the waypoints part of the state, (ii) introducing binary variables to enforce the constraint of passing once through each waypoint, and (iii) applying a fifth-degree Gauss-Lobatto direct collocation method to tackle the dynamic constraints. High degree interpolation polynomials allow the number of variables of the problem to be reduced for a given numerical precision. Finally, the problem will became a NLP problem with non-convex feasible region as far as we fix the sequence of waypoints. And the solution for this NLP problem is achieved in IPOPT solver in a short time. Further explanation about the optimal control problem transformations are in [3].

4   Numerical Experiments

In this section, the results of several numerical experiments where the robot is constraint to pass through the waypoints listed in Table 1 for the testbed lines (LIN) and listed in Table 2 for the testbed lattice (LAT).

Table 1.   Coordinates of the waypoints used in the experiments LIN

$p_1 = (0.455718, 0.660622)$	$p_2 = (0.472266, 0.616427)$	$p_3 = (0.510878, 0.513305)$
$p_4 = (0.538458, 0.439647)$	$p_5 = (0.571554, 0.351256)$	$p_6 = (0.610167, 0.248135)$
$p_7 = (0.335096, 0.591699)$	$p_8 = (0.450359, 0.571340)$	$p_9 = (0.536806, 0.556071)$
$p_{10} = (0.623253, 0.540801)$	$p_{11} = (0.767332, 0.515353)$	$p_{12} = (0.911410, 0.489904)$
$p_{13} = (0.331795, 0.685981)$	$p_{14} = (0.397159, 0.636696)$	$p_{15} = (0.527889, 0.538125)$
$p_{16} = (0.658618, 0.439554)$	$p_{17} = (0.789347, 0.340984)$	$p_{18} = (0.854711, 0.291699)$

Table 2.   Coordinates of the waypoints used in the experiments LAT

$p_1 = (-0.181751, 0.581213)$	$p_2 = (-0.111041, 0.510503)$	$p_3 = (-0.0403301, 0.439792)$
$p_4 = (0.0303806, 0.369081)$	$p_5 = (0.101091, 0.29837)$	$p_6 = (0.171802, 0.22766)$
$p_7 = (-0.111041, 0.651924)$	$p_8 = (-0.0403301, 0.581213)$	$p_9 = (0.0303806, 0.510503)$
$p_{10} = (0.101091, 0.439792)$	$p_{11} = (0.171802, 0.369081)$	$p_{12} = (0.242513, 0.29837)$
$p_{13} = (-0.04033, 0.722635)$	$p_{14} = (0.0303806, 0.651924)$	$p_{15} = (0.101091, 0.581213)$
$p_{16} = (0.171802, 0.510503)$	$p_{17} = (0.242513, 0.439792)$	$p_{18} = (0.313223, 0.369081)$

In Table 3 we report the results obtained while using the biased randomization in the *RR* problem. It has been implemented in a C++ code which generates a set of good discrete path where those with minimum total angle are evaluated in IPOPT optimization engine. For each instance we indicate the correspondent testbed and the number of waypoints considered. In all the experiments we use the initial time $t_I = 0$ [s] and the final times $t_F = 4$ [s] and $t_F = 6$ [s] for instances of 12 and 18 waypoints, respectively.

Also, instances with cases A, B and C differ in the location of the final point p_F which will be specified in each instance case.

Finally, Fig. 4 shows the lines (LIN) and lattice (LAT) configurations. Also, depicts the continuous path of minimum energy consumption in couple of instances, according to our Multi-Start approach.

Table 3. Multi-Start approach pseudo-optimal discrete paths

Instance	initial point and final point sequence of waypoints from Multi-Start

LIN-12A $p_I = (1.0843, 0.459365)$ $p_F = (0.637747, 0.174476)$
$(p_I, p_{12}, p_{11}, p_{10}, p_9, p_8, p_7, p_1, p_2, p_3, p_4, p_5, p_6, p_F)$

LIN-12B $p_I = (1.0843, 0.459365)$ $p_F = (0.2027, 0.6151)$
$(p_I, p_{12}, p_{11}, p_1, p_2, p_3, p_4, p_5, p_6, p_{10}, p_9, p_8, p_7, p_F)$

LIN-18A $p_I = (0.2664, 0.7353)$ $p_F = (0.637747, 0.174476)$
$(p_I, p_{13}, p_{14}, p_{15}, p_{16}, p_{17}, p_{18}, p_{12}, p_{11}, p_{10}, p_9, p_8, p_7, p_1, p_2, p_3, p_4,$
$p_5, p_6, p_F)$

LIN-18B $p_I = (0.2664, 0.7353)$ $p_F = (0.933149, 0.232556)$
$(p_I, p_{13}, p_7, p_9, p_{10}, p_{11}, p_{12}, p_6, p_5, p_4, p_3, p_2, p_1, p_{14}, p_8, p_{15}, p_{16},$
$p_{17}, p_{18}, p_F)$

LIN-18C $p_I = (0.2664, 0.7353)$ $p_F = (1.0843, 0.459365)$
$(p_I, p_{13}, p_{14}, p_{15}, p_{16}, p_{17}, p_{18}, p_6, p_5, p_4, p_3, p_2, p_1, p_7, p_8, p_9, p_{10},$
$p_{11}, p_{12}, p_F)$

LAT-12 $p_I = (1.0843, 0.459365)$ $p_F = (0.637747, 0.174476)$
$(p_I, p_1, p_8, p_9, p_{10}, p_{11}, p_{12}, p_6, p_5, p_4, p_3, p_2, p_7, p_F)$

LAT-18 $p_I = (1.0843, 0.459365)$ $p_F = (0.637747, 0.174476)$
$(p_I, p_1, p_8, p_9, p_{10}, p_{11}, p_{17}, p_{16}, p_{15}, p_{14}, p_{13}, p_7, p_2, p_3, p_4, p_5, p_6,$
$p_{12}, p_{18}, p_F)$

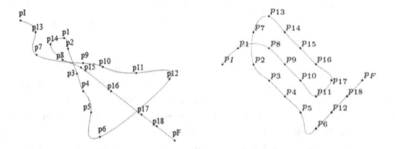

Fig. 4. Continuous path for instances LIN-18B and LAT-18 using biased randomization

In order to compare the previous work with the BONMIN solver and our Multi-Start approach we applied both methods in each of the instances in Table 3. Notice that any of these methodologies is an exact method, the first one is based on BB algorithm and ours is meta-heuristics approach, so the optimal solution not guarantee in any case.

Regarding to optimality, in both procedures we have obtained the same discrete path, except for LIN-18B and LAT-12A. And therefore, the same continuous trajectory and energy consumption returned from IPOPT, except for LIN-18B and LAT-12A. In LIN-18B the discrete path obtained in BONMIN is (p_I, p_{13}, p_1, p_3, p_4, p_5, p_6, p_{12}, p_{11}, p_{10}, p_9, p_8, p_7, p_{14}, p_2, p_{15}, p_{16}, p_{17}, $p_{18,}$ p_F) were the energy consumption is reduced from our solution 2,29 %. However, in LAT-12A the discrete path obtained in BONMIN is (p_I, p_1, p_2, p_7, p_3, p_4, p_5, p_6, p_{12}, p_{11}, p_{10}, p_9, p_8, p_F) were the energy consumption has increased from our solution more than 5 %.

Regarding to computational times, our Multi-Start approach provides us the solution in few minutes. In a similar CPU, the BONMIN solver is also able to solve lines instances (LIN) in few minutes, but dozen of hours for the lattice configuration. However, the latter is the realistic configuration that often occurs in the industrial robots.

5   Conclusions

This paper we study the motion planning problem where to face high complexities. In one hand the combinatorial problem, on the other hand the dynamic model. For the latter, most of the methods interpolation to determine velocity profile along the path, however the resulting trajectory can be dynamically unfeasible due to physical limitations of the actuators. In this work we proposed a Multi-Start approach which provides not only the discrete path but also ensure the dynamical feasibility of the trajectory. Therefore, we apply biased randomization to obtain a set of good discrete paths and the NLP problem to be displayed in IPOPT solver to obtain the dynamicity of the trajectory. And, since we evaluate continuous trajectories only if the discrete path is promising, the numerical experiments in terms of computational times are much lower than those in [3] where there is no discrimination in the use of IPOPT solver.

Acknowledgment. This work has been partially supported by the Spanish Ministry of Economy and Competitiveness (TRA2013-48180-C3-P and TRA2015-71883-REDT), FEDER, and the Catalan Government (2014-CTP-00001).

References

1. La Valle, S.: Planning Algorithms. Cambridge University Press, Cambridge (2006)
2. Siciliano, B., Khatib, O. (eds.): Handbook of Robotics. Springer, Heidelberg (2008)
3. Bonami, P., Olivares, A., Staffetti, E.: Energy-optimal multi-goal motion planning for planar robot manipulators. J. Optim. Theor. Appl. **163**(1), 80–104 (2014)

4. Bonami, P., Biegler, L.T., Conn, A.R., Cornuéjols, G., Grossmann, I.E., Laird, C.D., Lee, J., Lodi, A., Margot, F., Sawaya, N., Wächter, A.: An algorithmic framework for convex mixed integer nonlinear programs. Discrete Optim. 5(2), 186–204 (2008)
5. Pirnay, H., López-Negrete, R., Biegler, L.T.: sIPOPT Reference Manual. Carnegie Mellon University (2011). https://projects.coin-or.org/Ipopt
6. Juan, A., Faulin, J., Ruiz, R., Barrios, B., Caballé, S.: The SR-GCWS hybrid algorithm for solving the capacitated vehicle routing. Appl. Soft Comput. 10, 215–224 (2010)

Bonferroni Means with the Adequacy Coefficient and the Index of Maximum and Minimum Level

Fabio Blanco-Mesa[1(✉)] and José M. Merigó[2]

[1] Department of Business Administration, Faculty of Business Administration,
Antonio Nariño University, 7 Avenue, 21-84, 150001 Tunja, Colombia
fabio.blanco@uan.edu.co
[2] Department of Management Control and Information Systems,
School of Economic and Business, University of Chile,
Av. Diagonal Paraguay, 257, 8330015 Santiago, Chile

Abstract. The aim of the paper is to develop new aggregation operators using Bonferroni means, OWA operators and some distance and norms measures. We introduce the BON-OWAAC and BON-OWAIMAM operators. We are able to include adequacy coefficient and the maximum and minimum level in the same formulation with Bonferroni means and OWA operator. The main advantages on using these operators are that they allow considering continuous aggregations, multiple-comparison between each argument and distance measures in the same formulation. The numerical sample is focused on an entrepreneurial example in the sport industry in Colombia.

Keywords: Bonferroni means · OWA operators · Distance measures

1 Introduction

In the literature, there are a wide range of methods that allows aggregating information [1–6], which allow obtaining representative values of the aggregated information. One of the most popular models used is the ordered weighted averaging OWA operator [1], from which has developed a great deal of extension in combination with others mathematical models. One of these models are selection indices [3]. These are called ordered weighted averaging distance (OWAD) [3, 7], ordered weighted averaging adequacy coefficient (OWAAC) operator and ordered weighted averaging index of maximum and minimum level (OWAIMAM) operator [8, 9] and allow aggregating information through the comparison between two elements in order to get a representative value. Likewise, a new aggregation operator is proposed by using Bonferroni means (BM) [10], which allows making multiple-comparison between input arguments and capturing its interrelationship. Yager [11] combined OWA operator with BM proposing a new aggregation operator called BON-OWA and suggested a generalization of this operator. This new aggregation operator aroused the curiosity of the scientific community so that multiple authors study and develop new models based on it [12–19].

© Springer International Publishing Switzerland 2016
R. León et al. (Eds.): MS 2016, LNBIP 254, pp. 155–166, 2016.
DOI: 10.1007/978-3-319-40506-3_16

The aim of this paper is to develop new mathematical application based on Bonferroni means, OWA operator and some distance measure. This application consists in that BM in combination with OWAIMAN and OWAAC operator. The main advantage of this proposition is that it allows considering continuous aggregations, multiple-comparison between each argument and distance measures in the same formulation. The structure of this paper is as follows: In Sect. 2, basic concepts Bonferroni means, OWA operators and distance measures are briefly reviewed and new proposition is presented. In Sect. 3, new method based on adequacy coefficient and in combination with BON-OWAAC and BON-OWAIMAN is presented. In Sect. 4, summary and main conclusion are presented.

2 Preliminaries

In this section, we briefly review of Bonferroni means, OWA operator, BON-OWA, distance measures and OWAD in order to develop new tools based on distance measures in combination with Bonferroni means and OWA operators.

2.1 Bonferroni Means

The Bonferroni mean [10] is another type of mean that can be used in the aggregation process in order to present the information. Recently several authors have used it with OWA operators [11, 12], uncertain [20], linguistic variables [17, 21], intuitionistic [13, 22] and hesitant representation [14, 23]. It can be defined by using the following expression.

$$B(a_1, a_2, \ldots, a_n) = \left(\frac{1}{n} \frac{1}{1-n} \sum_{\substack{j=1 \\ j \neq k}}^{n} a_j^q \right)^{\frac{1}{r+q}}, \tag{1}$$

where r and q are parameters such that $r, q \geq 0$ and the arguments $a \geq 0$. By rearranging the terms (Yager 2009), it can be also formulate in the following way:

$$B(a_1, a_2, \ldots, a_n) = \left(\sum_{k=1}^{n} a_k^r \left(\frac{1}{1-n} \sum_{\substack{j=1 \\ j \neq k}}^{n} a_j^q \right) \right)^{\frac{1}{r+q}}. \tag{2}$$

2.2 OWA Operators

2.2.1 The OWA Operator

The OWA operator [1] provides a parameterized class of mean type of aggregation operators. It can be defined as follows.

Definition 1. An OWA operator of dimension n is a mapping OWA : $R^n \to R$ that has an associated weighting vector W of dimension n with $w_j \in [0, 1]$ and $\sum_{j=1}^{n} w_j = 1$, such that:

$$OWA(a_1, a_2, \ldots, a_n) = \sum_{j=1}^{n} w_j b_j, \qquad (3)$$

where b_i is the jth largest of the a_i.

2.2.2 Bonferroni OWA

The Bonferroni OWA [11] is mean type aggregation operator. It can be defined by using the following expression.

$$BON - OWA(a_1, \ldots, a_n) = \left(\frac{1}{n} \sum_i a_i^r OWA_W(V^i) \right)^{\frac{1}{r+q}}, \qquad (4)$$

where $OWA_W(V^i) = \left(\begin{array}{c} \frac{1}{n-1} \sum_{\substack{j=1 \\ j \neq i}}^{n} a_j^q \end{array} \right)$ with (V^i) being the vector of all a_j except a_i and w being an $n - 1$ vector W_i associated with α_i whose components w_{ij} are the OWA weights. Let W be an OWA weighting vector of dimension $n - 1$ with components $w_i \in [0, 1]$ when $\sum_i w_i = 1$. Then, we can define this aggregation as $OWA_W(V^i) = \left(\sum_{j=1}^{n-1} w_i a_{\pi_k(j)} \right)$, where $a_{\pi_k(j)}$ is the largest element in the tuple V^i and $w_i = \frac{1}{n-1}$ for all i. Thus, we have observed that this aggregation is equal at the original case. Furthermore, according to [11] the weight vector w_i can be stipulated by different ways. One approach is to directly specify the vector W. Other form is using [24] approach $- \sum_{j=1}^{n-1} w_j \ln(w_j)$ such as $\sum_{j=1}^{n-1} w_j \frac{n-j}{n-1} = \alpha$, $\sum_{j=1}^{n-1} w_j = 1$, $0 \leq w_j \leq 1$. Another approach is via BUM function f, in which we get $w_j = f\left(\frac{j}{n-1}\right) - f\left(\frac{j-1}{n-1}\right)$. Based on this method is develop other approach, which starts with a parameterized family of BUM functions and define the desired aggregation by specifying the value associated parameter [11]. Another parameter function is $f(x) = x^r$ for $r > 0$, where from r we get a particular function. Attitudinal character is such that $\alpha = \frac{1}{r+1}$ and if we specify α we can obtain r $= \frac{1-\alpha}{\alpha}$ [11].

2.3 Distance Measures

2.3.1 The Hamming Distance

The Hamming distance [25] is a useful technique for calculating the differences between two elements, two sets, etc. In fuzzy set theory, it can be useful, for example, for the calculation of distances between fuzzy sets, interval-valued fuzzy sets, intuitionistic fuzzy sets and interval-valued intuitionistic fuzzy sets. For two sets A and B, the weighted Hamming distance can be defined as follows.

Definition 2. A weighted Hamming distance of dimension n is a mapping d_{WH} : $R^n x R^n \to R$ that has an associated weighting vector W of dimension n with the sum of the weights being 1 and $w_j \in [0, 1]$ such that:

$$d_{WH}(\langle x_1, y_1 \rangle, \ldots, \langle x_n, y_n \rangle) = \sum_{j=1}^{n} w_j |x_i - y_i|, \tag{5}$$

where x_i and y_i are the ith arguments of the sets X and Y.

2.3.2   The Adequacy Coefficient

The adequacy coefficient [26, 27] is an index used for calculating the differences between two elements, two sets, etc. It is very similar to the Hamming distance with the difference that it neutralizes the result when the comparison shows that the real element is higher than the ideal one. For two sets A and B, the weighted adequacy coefficient can be defined as follows.

Definition 3. A weighted adequacy coefficient of dimension n is a mapping K : $[0, 1]^n x [0, 1]^n \to [0, 1]$ that has an associated weighting vector W of dimension n with the sum of the weights 1 and $w_j \in [0, 1]$ such that:

$$K(\langle x_1, y_1 \rangle, \ldots, \langle x_n, y_n \rangle) = \sum_{i=1}^{n} w_i [1 \wedge (1 - x_i + y_i)] \tag{6}$$

where x_i and y_i are the ith arguments of the sets X and Y.

2.3.3   The Index of Maximum and Minimum Level

The index of maximum and minimum level is an index that unifies the Hamming distance and the adequacy coefficient in the same formulation [28]. For two sets A and B, the weighted index of maximum and minimum level can be defined as follows.

Definition 4. An AWIMAM of dimension n is a mapping $K : [0, 1]^n x [0, 1]^n \to [0, 1]$ that has an associated weighting vector W of dimension n with the sum of the weights 1 and $w_j \in [0, 1]$ such that:

$$\eta(\langle x_1, y_1 \rangle, \ldots, \langle x_n, y_n \rangle) = \sum_u Z_i(u) * |x_i(u) - y_i(u)| + \sum_v Z_i(v) * [0 \vee x_i(v) - y_i(v)], \tag{7}$$

where x_i and y_i are the *ith* arguments of the sets X and Y.

2.4   OWA Operator and Distance Measures

The OWAD operator [3, 7] is an aggregation operator that uses OWA operators and distance measures in the same formulation. It can be defined as follows for two sets X and Y.

Definition 5. An OWAD operator of dimension n is a mapping $OWAD : R^n x R^n \rightarrow R$ that has an associated weighting vector W, $\sum_{j=1}^{n} w_j = 1$ and $w_j \in [0, 1]$ such that:

$$OWAD(\langle x_1, y_1 \rangle, \ldots, \langle x_n, y_n \rangle) = \sum_{j=1}^{n} w_j D_j, \qquad (8)$$

where D_j represents the *jth* largest of the $|x_i - y_i|$.

The OWAAC operator [6, 7, 29] is an aggregation operator that uses the adequacy coefficient and the OWA operator in the same formulation. It can be defined as follows for two sets X and Y.

Definition 6. An OWAAC operator of dimension n is a mapping $OWAAC : [0, 1]^n x [0, 1]^n \rightarrow [0, 1]$ that has an associated weighting vector W, with $w_j \in [0, 1]$ and $\sum_{j=1}^{n} w_j = 1$, such that:

$$OWAAC(\langle x_1, y_1 \rangle, \ldots, \langle x_n, y_n \rangle) = \sum_{j=1}^{n} w_j K_j \qquad (9)$$

where K_j represents the jth largest of $[1 \wedge (1 - x_i + y_i)]$, $[1 \wedge (1 - x_i + y_i)]$.

The OWAIMAM operator [3, 5, 30] is an aggregation operator that uses the Hamming distance, the adequacy coefficient and the OWA operator in the same formulation. It can be defined as follows.

Definition 7. An OWAIMAM operator of dimension n, is a mapping $OWAIMAM : [0, 1]^n x [0, 1]^n \rightarrow [0, 1]$ that has an associated weighting vector W, with $w_j \in [0, 1]$ and the sum of the weights is equal to 1, such that:

$$OWAIMAM(\langle x_1, y_1 \rangle, \langle x_2, y_2 \rangle, \ldots, \langle x_n, y_n \rangle) = \sum_{j=1}^{n} w_j K_j, \qquad (10)$$

where K_j represents the *jth* largest of all the $|x_i - y_i|$ and the $[0 \vee (x_i, y_i)]$.

2.5 Bonferroni Means and Distance Measures

In this section, we briefly review of aggregation operators such as: Bonferroni distance (BD), Bonferroni OWAD (BON-OWAD). Likewise, we present news aggregation operators using Bonferroni means, OWA operator Adequacy coefficient and the index of maximum and minimum level.

2.5.1 Bonferroni Distance

Bonferroni distance using Bonferroni means, distance measure and OWA operator. This proposal has suggested a group of operators, such as: Bonferroni distance (BD), Bonferroni OWAD. For this study, we have focus on BD and BON-OWAD concepts, which are defined as follows.

Definition 8. Bonferroni distance for two sets $X = \{x_1, x_2, \ldots, x_n\}$ and $Y = \{y_1, y_2, \ldots, y_n\}$ is given by:

$$BD(\langle x_1, y_1 \rangle, \ldots, \langle x_n, y_n \rangle) = \left(\frac{1}{n} \sum\nolimits_{k=1}^{n} d_i^r \left(\frac{1}{n-1} \sum\nolimits_{\substack{j=1 \\ j \neq k}}^{n} d_j^q \right) \right)^{\frac{1}{r+q}}, \qquad (11)$$

where d_i and d_j are the individual such that $d_i = |x_i - y_i|$ and $d_j = |x_j - y_j|$.

Definition 9. A BON-OWAD distance for two sets. $X = \{x_1, x_2, x_3 \ldots x_n\}$ and $Y = \{y_1, y_2, \ldots, y_n\}$ is given by:

$$BON - OWAD(\langle x_1, y_1 \rangle, \ldots, \langle x_n, y_n \rangle) = \left(\frac{1}{n} \sum\nolimits_i D_i^r OWAD_{\omega_i}(V^i) \right)^{\frac{1}{r+q}}, \qquad (12)$$

where $OWAD_{\omega_i}(V^i) = \left(\frac{1}{n-1} \sum\nolimits_{\substack{j=1 \\ j \neq i}}^{n} D_j^q \right)$ with (V^i) being the vector of all $|x_j - y_j|$

except $|x_i - y_i|$ and ω_i being an $n-1$ vector W_i associated with α_i whose components w_{ij} are the OWA weights. Likewise, D_i is the kth smallest of the individual distance $|x_i - y_i|$.

3 New Method Based on the Adequacy Coefficient and the Index of Maximum and Minimum Level in Combination with BON-OWA

The adequacy coefficient was proposed by [26, 27] and OWAAC operator was proposed by [6, 7, 29], the combination of both allow aggregating information through the comparison between two elements with the characteristic that it neutralizes the result when the comparison shows that the real element is higher than the ideal one.

Proposition 1. Bonferroni adequacy coefficient for two sets $X = \{x_1, x_2, \ldots, x_n\}$ and $Y = \{y_1, y_2, \ldots, y_n\}$ is given by:

$$BAC(\langle x_1, y_1 \rangle, \ldots, \langle x_n, y_n \rangle) = \left(\sum\nolimits_{k=1}^{n} d_k^r \left(\frac{1}{1-n} \sum\nolimits_{\substack{j=1 \\ j \neq k}}^{n} d_j^q \right) \right)^{\frac{1}{r+q}}, \qquad (13)$$

where d_i and d_j are the individual such that $d_i = [1 \wedge (1 - x_i + y_i)]$ and $d_j = [1 \wedge (1 - x_j + y_j)]$.

Proposition 2. A BON-OWAAC distance for two sets $X = \{x_1, x_2, \ldots, x_n\}$ and $Y = \{y_1, y_2, \ldots, y_n\}$ is given by:

$$BON - OWAAC(\langle x_1, y_1 \rangle, \ldots, \langle x_n, y_n \rangle) = \left(\frac{1}{n} \sum\nolimits_{k=1}^{n} D_k^r OWAAC_{\omega_i}(V^i) \right)^{\frac{1}{r+q}}, \quad (14)$$

where $OWAAC_{\omega_i}(V^i) = \left(\frac{1}{1-n} \sum\limits_{\substack{j=1 \\ j \neq k}}^{n} D_j^q \right)$ with (V^i) being the vector of all $1 \wedge$

$(1 - x_j + y_j)$ except $1 \wedge (1 - x_i + y_i)$ and ω_i being an $n-1$ vector W_i associated with α_i whose components w_{ij} are the OWA weights. Likewise, D_i is the kth smallest of the individual distance $[1 \wedge (1 - x_i + y_i)]$.

The index of maximum and minimum level was proposed by [28] and OWAIMAN was proposed by [3, 5, 30] the combination of both allow aggregating information through the comparison between two elements with the characteristic. Besides, IMAM index unifies the Hamming distance and the adequacy coefficient allowing to have the characteristic in the same formulation.

Proposition 3. Bonferroni index of maximum and minimum level for two sets $X = \{x_1, x_2, \ldots, x_n\}$ and $Y = \{y_1, y_2, \ldots, y_n\}$ is given by:

$$BIMAM(\langle x_1, y_1 \rangle, \ldots, \langle x_n, y_n \rangle) = \left(\sum\nolimits_{k=1}^{n} d_k^r \left(\frac{1}{1-n} \sum\limits_{\substack{j=1 \\ j \neq k}}^{n} d_j^q \right) \right)^{\frac{1}{r+q}}, \quad (15)$$

where d_i and d_j are the individual such that $d_i = |x_i - y_i|$ and the $[0 \vee (x_i, y_i)]$ and $d_j = [|x_j - y_j|$ and the $[0 \vee (x_j, y_j)]$.

Proposition 4. A BON-OWAIMAM distance for two sets $X = \{x_1, x_2, \ldots, x_n\}$ and $Y = \{y_1, y_2, \ldots, y_n\}$ is given by:

$$BON - OWAIMAM(\langle x_1, y_1 \rangle, \ldots, \langle x_n, y_n \rangle) = \left(\frac{1}{n} \sum\nolimits_{k=1}^{n} D_k^r OWAIMAM_{\omega_i}(V^i) \right)^{\frac{1}{r+q}},$$

$$(16)$$

where $OWAIMAM_{\omega_i}(V^i) = \left(\frac{1}{1-n} \sum\limits_{\substack{j=1 \\ j \neq k}}^{n} D_j^q \right)$ with (V^i) being the vector of all

$[|x_j - y_j|$ and the $[0 \vee (x_j, y_j)]$ except $|x_i - y_i|$ and the $[0 \vee (x_i, y_i)]$ and ω_i being an $n-1$ vector W_i associated with α_i whose components w_{ij} are the OWA weights. Likewise, D_i is the kth smallest of the individual distance $|x_i - y_i|$ and the $[0 \vee (x_i, y_i)]$.

These OWA operators are commutative, monotonic, non-negative and reflexive. They are commutative from the OWA perspective because $f(\langle x_1, y_1 \rangle, \ldots, \langle x_n, y_n \rangle) = f(\langle c_1, d_1 \rangle, \ldots, \langle c_n, d_n \rangle)$ where $(\langle x_1, y_1 \rangle, \ldots, \langle x_n, y_n \rangle)$ is any permutation of the arguments $(\langle c_1, d_1 \rangle, \ldots, \langle c_n, d_n \rangle)$. They are also commutative from the distance measure perspective because $f(\langle x_1, y_1 \rangle, \ldots, \langle x_n, y_n \rangle) = f(\langle y_1, x_1 \rangle, \ldots, \langle y_n, x_n \rangle)$. They are monotonic because if $|x_i - y_i| \geq |c_i - d_i|$, for all i, then $f(\langle x_1, y_1\rangle,\ldots, \langle x_n, y_n\rangle) \geq f(\langle c_1, d_1\rangle,$ $\ldots, \langle c_n, d_n\rangle)$. Non-negativity is also accomplished always, that is, $f(\langle x_1, y_1\rangle,\ldots, \langle x_n, y_n\rangle)$ ≥ 0. Finally, they are also reflexive because $f(\langle x_1, x_1\rangle,\ldots, \langle x_n, x_n\rangle) = 0$.

Another issue to consider is the different measures used in the OWA literature for characterizing the weighting vector [31]. As mentioned above, weighting vector can be stipulated by numbers ways. Hence, we consider the entropy of dispersion, the balance operator, the divergence of W and the degree of orness [1, 31]. The entropy of dispersion is defined as follows:

$$H(W) = -\left(\frac{1}{n}\sum_i \ln(w_i)\left(\sum_{\substack{j=1\\j\neq i}}^n w_i \ln(w_i)\right)\right)^{\frac{1}{r+q}}. \tag{17}$$

For the balance operator, we get

$$Bal(W) = \left(\frac{1}{n}\sum_{i=1}^n \left(\frac{n+1-2_i}{n-1}\right)\left(\sum_{\substack{j=1\\j\neq i}}^n \left(\frac{n+1-2_j}{n-1}\right)w_i\right)\right)^{\frac{1}{r+q}}. \tag{18}$$

For the divergence of W, we get

$$Div(W) = \left(\frac{1}{n}\sum_{i=1}^n \left(\frac{n-i}{n-1} - \alpha(W)\right)^2\left(\sum_{\substack{j=1\\j\neq i}}^n w_i\left(\frac{n-j}{n-1} - \alpha(W)\right)^2\right)\right)^{\frac{1}{r+q}}. \tag{19}$$

For the degree of orness, we get

$$\alpha(W) = \left(\frac{1}{n}\sum_{i=1}^n \left(\frac{n-i}{n-1}\right)\left(\sum_{\substack{j=1\\j\neq i}}^n w_i\left(\frac{n-j}{n-1}\right)\right)\right)^{\frac{1}{r+q}}. \tag{20}$$

In order to understand the BON-OWAAC and BON-OWAIMAM numerically, lets present a simple example. The numerical sample is focused on an entrepreneurial example in the sport industry in Colombia. We have assumed that entrepreneurs potential want to select an industry to start a business within sports sector and there are six possible options NB_1 (textile and manufacturing industry), NB_2 (sports services), NB_3 (sports and leisure events), NB_4 (sports management consultancy services), NB_5 (media) and NB_6 (marketing, selling and distributions). It has fixed the *"ideal level"* these are current conditions available under the environment to start a business in Colombia. Entrepreneurial framework conditions (EFC's) in Colombia defines this level: C_1: 0,46; C_2: 0,56; C_3: 0,52; C_4: 0,6; C_5: 0,46; C_6: 0,64; C_7: 0,48; C_8: 0,56; C_9: 0,58; C_{10}: 0,56; C_{11}: 0,66; C_{12}: 0,62. It has fixed the "real level" of each characteristic for all the different sectors into the considered sports industry. It is also remarkable that the real level is identified as "necessary conditions" (see Table 1).

In this application, it assumes that the experts decide to take into account the most problematic factor for doing business rates in Colombia as weighting vector W_{PF}. Global Competitiveness Report, 2012, defines this vector. W_{PF1}: 0.0; W_{PF2}: 0,277; W_{PF3}: 0,396; W_{PF4}: 0,599; W_{PF5}: 0,599; W_{PF6}: 0,619; W_{PF7}: 0,634; W_{PF8}: 0,688; W_{PF9}: 0,837; W_{PF10}: 0,861; W_{PF11}: 0,881 and W_{PF12}: 0,886. That resulting data has been Normalized (N) to establish the weight of each factor and Inverse Normalized (IN) for showing positive factor effects.

Table 1. Necessary conditions to start a new business in sport industry

Y	C_1	C_2	C_3	C_4	C_5	C_6	C_7	C_8	C_9	C_{10}	C_{11}	C_{12}
NB_1	0,1	0,6	0,4	0,8	0,4	0,7	0,7	0,8	0,7	0,7	0,9	0,9
NB_2	0,4	0,6	0,4	0,2	0,7	0,6	0,3	0,6	0,6	0,6	0,7	0,6
NB_3	0,6	0,6	0,5	0,5	0,7	0,6	0,3	0,6	0,1	0,3	0,9	0,9
NB_4	0,1	0,6	0,5	0,3	0,7	0,9	0,2	0,2	0,5	0,7	0,5	0,3
NB_5	0,1	0,6	0,7	0,2	0,3	0,6	0,3	0,4	0,1	0,1	0,7	0,6
NB_6	0,1	0,6	0,2	0,2	0,6	0,6	0,5	0,8	0,6	0,7	0,7	0,9

The following shows the main results of the application. We have used several models in order to compare and show the versatility of the new proposition (see Table 2). It is noted that results change according to the algorithm used. These results are not similar but being ordered can coincide the number of ranking. However, it should be noted the importance of each algorithm and capabilities offered. Thus, BON-OWA algorithms combine the main characteristics of HD, AC, IMAM and OWA with BM showing the importance and interrelationships of each distance.

Table 2. Comparison with different models in order to start a new business in sports industry

Ind.	HD	AC	OWAD	OWAAC	OWA-IMAM	BD	BAC	BIMAM	BON-OWAD	BON-OWAAC	BON-OWAIMAM
NB_1	0,171	0,967	0,111	0,636	0,406	0,094	0,978	0,752	0,163	0,940	0,738
NB_2	0,095	0,939	0,058	0,604	0,364	0,077	0,943	0,718	0,135	0,828	0,618
NB_3	0,162	0,928	0,123	0,578	0,399	0,090	0,954	0,737	0,161	0,883	0,693
NB_4	0,181	0,870	0,115	0,558	0,329	0,122	0,905	0,694	0,221	0,804	0,612
NB_5	0,169	0,845	0,108	0,537	0,337	0,106	0,898	0,711	0,199	0,800	0,622
NB_6	0,165	0,911	0,094	0,606	0,350	0,103	0,949	0,671	0,191	0,901	0,689

In Table 3, it is shown the order of each option to start a new business in sport industry according to environmental factors in Colombia. In the first position, it is found alternative 1 and 2, which are the most favourable alternative for starting a new business. In the second position, it is found alternatives 3 and 6, which are far more likely to be developed. In the last position, it is found alternatives 4 and 5, are far less likely to lead to any entrepreneurial initiative.

Table 3. Order of each alternative to start a new business in sports industry

Ran	HD	AC	OWAD	OWAAC	OWAIMAM	BD	BAC	BIMAM	BON-OWAD	BON-OWAAC	BON-OWAIMAM
1	NB_2	NB_1	NB_2	NB_1	NB_1	NB_2	NB_1	NB_1	NB_2	NB_1	NB_1
2	NB_3	NB_2	NB_6	NB_6	NB_3	NB_3	NB_3	NB_3	NB_3	NB_6	NB_3
3	NB_6	NB_3	NB_5	NB_2	NB_2	NB_1	NB_6	NB_2	NB_1	NB_3	NB_6
4	NB_5	NB_6	NB_4	NB_3	NB_6	NB_6	NB_2	NB_5	NB_6	NB_2	NB_5
5	NB_1	NB_4	NB_1	NB_4	NB_5	NB_5	NB_4	NB_4	NB_5	NB_4	NB_2
6	NB_4	NB_5	NB_3	NB_5	NB_4	NB_4	NB_5	NB_6	NB_4	NB_5	NB_4

4 Conclusions

We have studied OWA operators, some distance measures and Bonferroni means in order to propose new aggregation operators. We have introduced new aggregation operators using AC and IMAM in the same formulation with Bonferroni means and OWA operator. The methods introduced are called BON-OWAAC and BON-OWAIMAM. The main advantages on using these operators are that they allow considering continuous aggregations, multiple-comparison between each argument and distance measures in the same formulation. Besides, each method has specific advantage. For BON-OWAAC the differences between two sets is established a threshold in the comparison process when one set is higher than the other so the results are equal from this point. For BON-OWAIMAM the differences between two sets is established using characteristics of both HD and AC in the same formulation. Likewise, we have obtained other methods such as BAC and BIMAM. Thus, we get a new group of distance family, which allows analysing the importance and interrelationship of each distance to be analysed. We have highlighted the versatility of these algorithms via studying within the complex relations, which is focused on selecting and ordering alternatives according to the subjective preferences of the decision-maker and the information available. The main implications by using both operators are that they can help to interpret social and economic conditions giving a holistic view of the environment. Thus, these algorithms enables interrelate all variables considered in the comparison. In future research, we will extend these new approaches to frameworks with induced aggregation operators, weighted information and moving averages. We will also study the applicability of these new aggregation operators in decision-making problems. Particularly, we will focus on the creation of creative and productive groups, interrelationship between entrepreneurs in co-working space and teamwork in business.

Acknowledgements. We are grateful with the Antonio Nariño University funds the publication of this work.

References

1. Yager, R.R.: On ordered weighted averaging aggregation operators in multicriteria decision-making. IEEE Trans. Syst. Man. Cybern. **18**, 183–190 (1988)
2. Herrera, F., Herrera-Viedma, E., Verdegay, J.L.: Direct approach processes in group decision making using linguistic owa operators. Fuzzy Sets Syst. **79**, 175–190 (1996)
3. Merigó, J.M.: Nuevas Extensiones a los Operadores OWA y su Aplicación en los Métodos de Decisión. Ph.D. Thesis, University of Barcelona (Spain) (2009). http://diposit.ub.edu/dspace/handle/2445/35378
4. Merigó, J.M., Gil-Lafuente, A.M.: The induced generalized OWA operator. Inf. Sci. **179**, 729–741 (2009)
5. Merigó, J.M., Gil-Lafuente, A.M., Gi-Aluja, J.: A new aggregation method for strategic decision making and its application in assignment theory. African J. Bus. Manage. **5**, 4033–4043 (2011)
6. Merigó, J.M., Gil-Lafuente, A.M.: The generalized adequacy coefficient and its application in strategic decision making. Fuzzy Econ. Rev. **13**, 17–36 (2008)
7. Merigó, J.M., Gil-Lafuente, A.M.: New decision-making techniques and their application in the selection of financial products. Inf. Sci. **180**, 2085–2094 (2010)
8. Merigó, J.M.: Fuzzy decision making with immediate probabilities. Comput. Ind. Eng. **58**, 651–657 (2010)
9. Merigó, J.M., Gil-Lafuente, A.M.: Decision making techniques with similarity measures and OWA operators. SORT – Stat. Oper. Res. Trans. **36**, 81–102 (2012)
10. Bonferroni, C.: Sulle medie multiple di potenze. Boll. dell'Unione Mat. Ital. **5**, 267–270 (1950)
11. Yager, R.R.: On generalized Bonferroni mean operators for multi-criteria aggregation. Int. J. Approx. Reason. **50**, 1279–1286 (2009)
12. Beliakov, G., James, S., Mordelová, J., Rückschlossová, T., Yager, R.R.: Generalized Bonferroni mean operators in multi-criteria aggregation. Fuzzy Sets Syst. **161**, 2227–2242 (2010)
13. Xu, Z., Yager, R.R.: Intuitionistic fuzzy Bonferroni means. IEEE Trans. Syst. Man Cybern. B Cybern. **41**, 568–578 (2011)
14. Zhu, B., Xu, Z., Xia, M.: Hesitant fuzzy geometric Bonferroni means. Inf. Sci. **205**, 72–85 (2012)
15. Beliakov, G., James, S.: On extending generalized Bonferroni means to Atanassov orthopairs in decision making contexts. Fuzzy Sets Syst. **211**, 84–98 (2013)
16. He, Y., He, Z., Wang, G., Chen, H.: Hesitant fuzzy power Bonferroni means and their application to multiple attribute decision making. IEEE Trans. Fuzzy Syst. 1–1 (2014)
17. Wei, G., Zhao, X., Lin, R., Wang, H.: Uncertain linguistic Bonferroni mean operators and their application to multiple attribute decision making. Appl. Math. Model. **37**, 5277–5285 (2013)
18. Park, J.H., Park, J.E.: Generalized fuzzy Bonferroni harmonic mean operators and their applications in group decision making. J. Appl. Math. **2013**, 14 (2013)
19. Xia, M., Xu, Z., Zhu, B.: Geometric Bonferroni means with their application in multi-criteria decision making. Knowledge-Based Syst. **40**, 88–100 (2013)
20. Xu, Z.: Uncertain Bonferroni mean operators. Int. J. Comput. Intell. Syst. (2012)
21. Li, D., Zeng, W., Li, J.: Note on uncertain linguistic Bonferroni mean operators and their application to multiple attribute decision making. Appl. Math. Model. **39**, 894–900 (2015)
22. Xia, M., Xu, Z., Zhu, B.: Generalized intuitionistic fuzzy Bonferroni means. Int. J. Intell. Syst. **27**, 23–47 (2012)

23. Zhu, B., Xu, Z.S.: Hesitant fuzzy Bonferroni means for multi-criteria decision making. J. Oper. Res. Soc. **64**, 1831–1840 (2013)
24. O'Hagan, M.: Using maximum entropy-ordered weighted averaging to construct a fuzzy neuron. In: Proceedings of the 24th Annual IEEE Asilomar Conference on Signals, Systems and Computers, pp. 618–623. Pacific Grove (1990)
25. Hamming, R.W.: Error detecting and error correcting codes. Bell Syst. Tech. J. **29**, 147–160 (1950)
26. Kaufmann, A., Gil-Aluja, J.: Introducción de la Teoría de los Subconjuntos Borrosos a la Gestión de las Empresas. Milladoiro, Santiago de Compostela (1986)
27. Kaufmann, A., Gil-Aluja, J.: Técnicas Operativas de Gestión para el Tratamiento de la Incertidumbre. Hispano Europea, Barcelona (1987)
28. Gil-Lafuente, J.: El Índice del Máximo y el Mínimo Nivel en la Optimización del Fichaje de un Deportista (2001)
29. Gil-Lafuente, A.M., Merigó, J.M.: On the use of the OWA operator in the adequacy coefficient. Model. Meas. Control D. **30**, 1–17 (2009)
30. Merigó, J.M., Gil-Lafuente, A.M.: Decision-making in sport management based on the OWA operator. Expert Syst. Appl. **38**, 10408–10413 (2011)
31. Merigó, J.M., Casanovas, M.: Decision-making with distance measures and induced aggregation operators. Comput. Ind. Eng. **60**, 66–76 (2011)

Minimizing Trigger Error in Parametric Earthquake Catastrophe Bonds via Statistical Approaches

Jesica de Armas[1](✉), Laura Calvet[1], Guillermo Franco[2],
Madeleine Lopeman[2], and Angel A. Juan[1]

[1] Computer Science Department,
Open University of Catalonia-IN3, Barcelona, Spain
{jde_armasa,lcalvetl,ajuanp}@uoc.edu
[2] Guy Carpenter & Company, LLC, Dublin, Ireland
{Guillermo.E.Franco,Madeleine.Lopeman}@guycarp.com

Abstract. The insurance and reinsurance industry, some governments, and private entities employ catastrophe (CAT) bonds to obtain coverage for large losses induced by earthquakes. These financial instruments are designed to transfer catastrophic risks to the capital markets. When an event occurs, a Post-Event Loss Calculation (PELC) process is initiated to determine the losses to the bond and the subsequent recoveries for the bond sponsor. Given certain event parameters such as magnitude of the earthquake and the location of its epicenter, the CAT bond may pay a fixed amount or not pay at all. This paper reviews two statistical techniques for classification of events in order to identify which should trigger bond payments based on a large sample of simulated earthquakes. These statistical techniques are effective, simple to interpret and to implement. A numerical experiment is performed to illustrate their use, and to facilitate a comparison with a previously published evolutionary computation algorithm.

Keywords: Catastrophe bonds · Risk of natural hazards · Classification techniques · Earthquakes · Insurance

1 Introduction

Strategies to provide coverage for large losses ensuing after earthquakes through parametric CAT bonds have been implemented since the 1990s. These financial instruments allow insurers, reinsurers, governments, private entities and catastrophe pools to cede risks of losses to the capital markets via a transparent mechanism that determines payments based on certain quantifiable event features. These instruments bypass the claims adjusting process and therefore can provide a very fast recovery of funds to their sponsor after an event. The principal of the bond can also be collateralized thus reducing default risk, and most importantly, their price has been relatively competitive versus traditional premiums. In addition, the capital markets offer great capacity, although capacity has not been in short supply in the (re)insurance market lately.

© Springer International Publishing Switzerland 2016
R. León et al. (Eds.): MS 2016, LNBIP 254, pp. 167–175, 2016.
DOI: 10.1007/978-3-319-40506-3_17

Parametric CAT bonds employ triggers (or algorithms) to determine the payment that should take place when an earthquake occurs. These triggers rely on obtainable physical characteristics of the event [1, 2] and since no parties can manipulate this information, the risk transfer mechanism is transparent and reduces moral hazard (the risk that the parties involved can influence the payment outcome).

Earthquakes around the world cause enormous losses, which are rarely insured. These financial impacts end up distorting people's livelihoods and national economies severely. Parametric instruments have the potential to reduce this problem by making earthquake insurance more accessible. Therefore, this work focuses on improving the algorithms that can increase the quality of these transactions.

The classical parameters used for earthquake risk transactions are the magnitude of the event, the location of the epicenter, and the depth of the hypocenter–the theoretical center of energy dispersion of an earthquake within our planet's crust.

Our main objective is to explore a few statistical techniques to automatically construct accurate triggers, i.e. triggers that induce payment when they should and that do not induce payment when they should not. A computational experiment is performed to analyze their performance, mainly in terms of accuracy and time required. In addition, our approach is compared with the metaheuristic-based one described in [4].

The remainder of this paper is organized as follows: Sect. 2 is devoted to describe the trigger mechanism in detail. Section 3 introduces two statistical techniques that can be applied. Afterwards, the experiments are performed in Sect. 4. Finally, Sect. 5 summarizes the main highlights of this paper.

2   The Trigger Mechanism

Consider a set of l earthquake events in a geographic region of interest A. An earthquake event i is characterized by a magnitude m_i, a hypocenter depth d_i, and epicenter coordinates (x_i, y_i) within A. A binary trigger will determine whether a payment is associated with event i. This response is represented by the variable B', whose values 1/0 indicate trigger/no-trigger (payment/no-payment). Two situations may arise: (1) at least one earthquake i triggers the bond ($B'_i = 1$) during its contract life, which means that the entire bond principal has to be disbursed and, as a consequence, the buyers of the bond lose their investment (the bond sponsors receive compensation), and (2) no earthquake triggers the bond during its life, in which case the principal is returned to the investors with interest.

Since the payment of a large sum of money is at stake, it is important that the trigger performs as desired, that is that the trigger responds positively to events that cause a large loss beyond a design threshold and that it does not respond for events that cause a loss below this threshold. The accuracy of the trigger determines its success in the market. Triggers that behave erratically erode the confidence of the markets in these tools and therefore jeopardize the risk transfer process. Hence the importance of designing accurate triggers that behave as they should.

To describe the accuracy of the trigger, first consider a reference variable B that represents the idealized behavior of the trigger, which depends on a measure based on the losses (typically monetary). For an earthquake event i, this variable can be described as follows:

$$B_i = \begin{cases} 0, & \text{if } L_i < L \\ 1, & \text{otherwise,} \end{cases}$$

where L_i is the actual loss caused and L is a loss threshold specified by the sponsor, usually expressed in terms of a specific return period. Events trigger this CAT bond only if the corresponding loss is above a given pre-specified threshold L.

The objective is to develop a mechanism that minimizes discrepancies between variables B and B' or the sum of errors $(E = \sum_{i=1}^{l} I(B_i = B'_i))$, or in other words, the lack of correlation between the output of the trigger and the ideal trigger.

A database including a set of events, their characteristics and the variable B for each event can be used to calculate trigger errors for this specific set of events. A measure of the loss has to be obtained or estimated to compute B. It is preferable to have a reliable historical dataset including a high number of events but in earthquake research this is not possible due to the low recurrence of earthquakes and the great uncertainty surrounding their associated losses. For this reason, the design of triggers for earthquake risk depend on the usage of earthquake risk models, simulators of earthquake losses of popular usage in the (re)insurance industry.

Note that the physical characteristics suggested above do not constitute the only options to design a parametric trigger. However, the fact that they tend to be easily available from respected third parties makes them suitable for this purpose.

According to the description offered in this section, the development of a trigger mechanism can be labeled as a *binary classification problem*, allowing us to employ a wide range of techniques to address it. In the following sections, some of them are introduced and tested, and their use is illustrated.

3 Our Approaches

Classification techniques [6] constitute a set of procedures from statistics and machine learning (more specifically, supervised learning) to determine a category or class for a given observation. Having a dataset of l observations composed of independent or explanatory variables $(X_1, X_2, \ldots, X_n)$, and a dependent or response variable Y, these techniques attempt to explain the relationships between the variables and/or classify new observations based on the values of the explanatory variables.

Nowadays, there are plenty of classification techniques. Some of the most employed, e.g. Linear Discriminant Analysis or Logistic Regression, have been applied for more than five decades. These are mainly linear methods. Boosted by the computing advances in the 1980s and 1990s, non-linear methods such as Classification Trees, Neural Networks and Support Vector Machines emerged and/or started to attract attention more recently.

Two well-known and powerful techniques to automatically design a trigger are presented here. The first one is Logistic Regression, which belongs to the most typical linear methods, while the second one is Classification Trees, which belongs to the non-linear methods. The reader interested in more extensive comprehensive and practical descriptions is referred to [5] and [7].

- **Logistic Regression** techniques are designed to model the posterior probabilities of each class by means of linear functions. These probabilities must be non-negative and sum to one.

$$P(Y = y | X_1 = x_1 \cap X_2 = x_2 \cap \ldots \cap X_n = x_n) = \frac{e^{\beta_0 + \beta_1 x_1 + \beta_2 x_2 + \ldots + \beta_n x_n}}{1 + e^{\beta_0 + \beta_1 x_1 + \beta_2 x_2 + \ldots + \beta_n x_n}}$$

These models are usually fit by maximum likelihood employing the Newton's method. The previous expression can be rewritten in terms of log-odds as follows:

$$\log\left(\frac{P(Y = y | X_1 = x_1 \cap X_2 = x_2 \cap \ldots \cap X_n = x_n)}{1 - P(Y = y | X_1 = x_1 \cap X_2 = x_2 \cap \ldots \cap X_n = x_n)}\right)$$
$$= \beta_0 + \beta_1 x_1 + \beta_2 x_2 + \ldots + \beta_n x_n$$

This technique is especially useful when the aim is to be able to explain (i.e., not only classify) the outcome based on the explanatory variables. Non-linear functions can be considered including interactions and transformations of the original variables.

- **Classification Trees**, contrary to global models (where a predictive formula is supposed to hold in the entire data space) such those of Logistic Regression, tries to partition the data space into small enough parts where a simple model can be applied. The results can be represented as a tree composed of internal and terminal (or leaf) nodes, and branches. Its non-leaf part is a procedure to determine for each observation which model (i.e., terminal node) will be used to classify it. At each internal node of the tree, the value of one explanatory variable is checked and, depending on the binary answer, the procedure continues to the left or to the right sub-branch. A classification is made when a leaf is reached.
  The most relevant advantage of this classifier is the easiness to understand what trees represent. They seem more closely mirror human decision-making than other techniques. Furthermore, trees require little data preparation, are able to handle both numerical and categorical data, and perform well (i.e., use standard computing resources in reasonable time) with large datasets.

Although most researchers focus on accuracy, in real-life applications many other characteristics may play an important role when comparing the solutions given by these techniques. Examples are popularity, easiness-to-implement or to-explain to non-experts, and existence of graphical representations or summaries of the outputs, among many others. Even assuming we are only interested in the accuracy, the best technique will depend on the data at hand. Consequently, we can only present a general discussion about the performance of these techniques.

Logistic Regression, as a regression approach, is a well-established technique, which enables the understanding of the effects of the explanatory variables on the response. Classification Trees constitute an efficient technique that only uses the most important variables, and results in a logic model. As other techniques studying non-linear relationships, they may easily overfit or underfit the model. Moreover, small changes in training data may lead to significant modifications. In addition, they may derive decisions that seem counterintuitive or are unexpected.

In conclusion, each technique has different characteristics that should be considered when addressing a classification problem.

4 Computational Experiments

This section illustrates the use of the techniques introduced, and compares the results with those obtained with the methodology proposed in [4]. The dataset analyzed is an earthquake catalog representing a sample of 10,000 years of seismicity, a total of 24,957 earthquakes, concerning Costa Rica. A more detailed description can be found in the aforementioned work.

As commented in Sect. 2, the construction of a trigger is driven by the minimization of the discrepancies between its outputs and those from a trigger with an idealized behavior. If the resulting trigger is expected to be useful for new or unseen observations, it is good to avoid employing the same observations for developing the trigger and assessing its performance. This could lead to a problem of overfitting (i.e., obtaining complex models that capture specificities of the data but do not generalize well for other observations). An effective and efficient technique to avoid this problem is to split the dataset into three different subsets: a training set used for constructing the triggers, a validation set employed to tune the parameters, and a test set required to assess their performance. We apply this approach considering the following weights, since they are the most typical used values: 50 %, 25 %, and 25 %, respectively.

The statistical experiment has been performed with R (version 2.15.0) [8], a freely available and widely used language and environment for statistical computing and graphics. The following packages have been utilized: *stats* for Logistic Regression and *rpart* for the Classification Trees. A confidence level of 0.95 is considered.

For each classifier, the corresponding confusion matrix (Table 1) is obtained, which summarizes the results. This matrix shows the number of times that a predicted value B' coincides or not with the real value B. Therefore, it is desirable that cells TP and TN contain high values (as high as the number of real triggers of each type), and cells FP and FN contain low values. The measured accuracy is employed to make comparisons. In addition, the sensitivity, the specificity, and the time required are reported. Both false positive and false negative are equally penalized, i.e. we only focus on minimizing the total number of errors.

Table 1. Structure of a confusion matrix

		Predicted Class	
		$B' = 0$	$B' = 1$
Actual class	$B = 0$	True Positive (TP)	False Positive (FP)
	$B = 1$	False Negative (FN)	True Negative (TN)

4.1   Logistic Regression

Initially, a complete model including all variables and the interactions among pairs is built. Variables xc and yc correspond to the coordinates where the earthquake happens. An approach based on backward elimination has been implemented to select the final model, which is represented in Table 2.

Table 2. Logistic Regression model

	Estimate	Standard error	z value	P-value
(Intercept)	−2954.62	377.19	−7.83	4.75e-15
m	13.74	15.39	0.89	0.37
d	−0.06	0.16	−0.37	0.71
xc	−36.92	4.71	−7.84	4.52e-15
yc	271.89	34.18	7.95	1.81e-15
m · d	0.09	0.02	5.29	1.22e-7
m · x	0.50	0.19	2.64	8.28e-3
m · y	3.04	0.56	5.43	5.79e-8
d · y	−0.07	0.01	−5.16	2.48e-7
x · y	3.43	0.43	8.00	1.22e-15

Accordingly, the probability of having to pay given the data observed for a new event is:

$$P(Y = 1 | M = m \cap D = d \cap XC = xc \cap YC = yc)$$
$$= \frac{e^{-2954.62 + 13.74m - 0.06d - 36.92xc + 271.89yc + 0.09m \cdot d + 0.50m \cdot x + 3.04m \cdot y - 0.07d \cdot y + 3.43x \cdot y}}{1 - 2954.62 + 13.74m - 0.06d - 36.92xc + 271.89yc + 0.09m \cdot d + 0.50m \cdot x + 3.04m \cdot y - 0.07d \cdot y + 3.43x \cdot y}$$

If this probability is above 0.5, the event is classified in the group 1; otherwise, it is assigned to group 0. The resulting accuracy is 0.995512. Table 3 shows the confusion matrix, where we can see good values for True Positives and Negatives, as desirable, but we also have 26 times in which we obtain a false negative.

Table 3. Confusion matrix

	B' = 0	B' = 1
B = 0	6210	2
B = 1	26	1

4.2   Classification Tree

In order to construct a Classification Tree, it is needed to specify the Complexity Parameter (a parameter to measure the tree cost-complexity). Values from 0.01 up to 0.20 have been tested. The best results correspond to the value 0.05.

Figure 1 shows the tree representation. The observations that satisfy the condition shown for each internal node go to the left, otherwise, they go to the right. The percentage shown at the bottom of each node indicates the proportion of observations

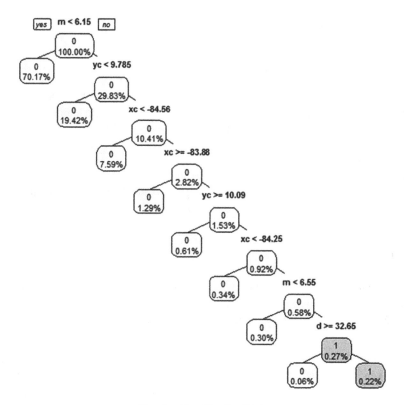

Fig. 1. Classification Tree

Table 4. Confusion matrix

	$B' = 0$	$B' = 1$
$B = 0$	6210	2
$B = 1$	13	14

that reach that node. The value above that percentage refers to the classification of the corresponding observations. The resulting accuracy is 0.997596 (see Table 4).

4.3 Comparative Analysis

In order to validate the application of these techniques for the development of triggers for earthquake catastrophe bonds, we compare our results with those provided in [4]. In the paper, the author proposes the construction of binary "cat-in-a-box" trigger mechanisms, where the geographical space is discretized in square boxes or sub-regions of the same size. Each sub-region belongs to a specific zone denoted as k. This constitutes a relatively simple and popular approach, where the aim is to set the parameters of a trigger mechanism for each zone by minimizing the trigger error. Concretely, the trigger mechanism has the following structure:

$$\forall (x_i, y_i) \in A_k, \quad B'_i = \begin{cases} 0, & \text{If } m_i < M_k \quad or \quad d_i > D_k \\ 1, & \text{If } m_i \geq M_k \quad or \quad d_i \leq D_k \end{cases}$$

where M_k and D_k are the parameters to set and represent the magnitude and depth thresholds, respectively. An Evolutionary Algorithm (EA) is implemented to address this problem and is executed for different combinations of box side lengths and number of zones. Although the paper does not report computational times, it is clear that the authors dedicated several hours to develop this ad-hoc proposal and to perform the parameter fine-tuning.

In order to compare our results with those published in [4], we use the same design process, assuming the entirety of the catalog is available to fine tune the trigger. Table 5 summarizes the results found with our techniques, and the EA. According to our results, one of our statistical techniques is able to obtain better performance than the EA in terms of accuracy and specificity. EA has a relatively good accuracy but it takes much longer in comparison. Notice that the sensitivity measures the proportion of positives that are correctly identified as such, the specificity measures the proportion of negatives that are correctly identified as such, and the accuracy measures the proportion of positives and negatives correctly identified as such.

Besides the computational time, statistical techniques present other relevant advantages regarding to:

- Scalability. While the structure of an EA has to be readjusted when more variables are taken into account, these techniques may be easily adapted to a larger parameter space. Being much faster, they may also work on bigger catalogs and still be able to provide results in a relatively short time.
- Implementation. There is a wide range of programs/programming languages that enable a free and simple implementation of these techniques such as R, Octave [3] or SciLab [9]. To facilitate their use further, most do propose default parameters for their algorithms or functions to perform an automatic parameter fine-tuning.
- Understanding. EAs usually rely on concepts from the field of biological evolution such as reproduction, mutation, recombination and selection, which may seem difficult to understand for non-expert users. On the other hand, the proposed techniques are presented as optimization problems where the aim is to minimize an error-based function considering some assumptions without using terms from fields other than mathematics, statistics and computer science.

Table 5. Results of the experiment comparing our approach and that described in [4]

	Accuracy	Sensitivity	Specificity	Time (s)
Classification trees	0.998237	0.998674	0.858974	0.34
Logistic regression	0.995713	0.996111	0.230769	0.64
EA [4]	0.998117	0.998794	0.804598	*

*Data not available.

5 Conclusions

Natural catastrophes may cause large economical losses that are currently underinsured, leaving large fractions of the population vulnerable to severe financial impacts. The insurance and reinsurance industry, governments and catastrophe pools have started to employ financial instruments such as parametric CAT bonds to cede these catastrophic risks to the capital markets. For this process to be satisfactory to all parties, it is essential that the triggers or algorithms we employ to determine payments are accurate and minimize errors.

We propose to address this trigger design process as a classification problem, employing well-known and powerful techniques from statistics. Our results show that employing logistic regression or classification trees, one can obtain results of equal or better accuracy than those published prior with the usage of evolutionary computation while also increasing the efficiency of the process. These observations point to the possibility that the state of the art in parametric risk transactions can be further enhanced by augmenting the parameter space used in the trigger design without further burdening the time required to design the trigger conditions.

Acknowledgments. This work has been partially supported by the Spanish Ministry of Economy and Competitiveness (TRA2013-48180-C3-P and TRA2015-71883-REDT), FEDER, and the Catalan Government (2014-CTP-00001).

References

1. Croson, D.C., Kunreuther, H.C.: Customizing reinsurance and cat bonds for natural hazards risks. In: Proceedings, Conference on Global Change and Catastrophic Risk Management, Laxenburg, Austria, IIASA, 6–9 June 1999
2. Cummins, J.D.: CAT bonds and other risk-linked securities: State of the market and recent developments, Social Science Research Network (2007). http://ssrn.com/abtract=105740
3. Eaton, J.W., et al.: GNU Octave (2015). http://www.octave.org
4. Franco, G.: Minimization of trigger error in cat-in-a-box parametric earthquake catastrophe bonds with an application to costa rica. Earthquake Spectra. **26**(4), 983–998 (2010)
5. Hastie, T., Tibshirani, R., Friedman, J.: The Elements of Statistical Learning, 2nd edn. Springer, New York (2009)
6. Kotsiantis, S.B.: Supervised machine learning: a review of classification techniques. Informatica **31**, 249–268 (2007)
7. Lantz, B.: Machine Learning. R. Packt Publishing, Birmingham (2013)
8. R Core Team: R: A Language and Environment for Statistical Computing. R Foundation for Statistical Computing. Vienna, Austria (2012). http://www.R-project.org/
9. Scilab Enterprises: Scilab: Free and Open Source software for numerical computation. Orsay, France (2012). http://www.scilab.org

Designing Teaching Strategies
with an Agent-Based Simulator

Iván García-Magariño[1(✉)], Inmaculada Plaza[2], Raúl Igual[3],
and Andrés S. Lombas[4]

[1] Department of Computer Science and Engineering of Systems,
University of Zaragoza, Teruel, Spain
ivangmg@unizar.es
[2] Department of Electronics Engineering and Communications,
University of Zaragoza, Teruel, Spain
inmap@unizar.es
[3] Department of Electrical Engineering, University of Zaragoza, Teruel, Spain
rigual@unizar.es
[4] Department of Psychology and Sociology, University of Zaragoza, Teruel, Spain
slombas@unizar.es

Abstract. In universities, a common task of lectures and professors is
to design an appropriate teaching strategy for each subject. The main
goal of this paper is to provide a mechanism that assists them in this
task. In particular, the current approach uses an agent-based simulator
for this purpose. This simulator allows teachers to simulate the influence
of a teaching strategy in the sociometric status of a specific group of
students. Since the cohesion of a group is usually related to its performance, teachers can choose between several possible strategies considering their simulated cohesion. This paper briefly presents two experiences
with the current approach in two different engineering grades (i.e. computer science engineering and electrical engineering) in the University of
Zaragoza.

Keywords: Agent-Based Simulator · Agent-oriented software engineering · Multi-agent system · Social simulation · Teaching strategy

1 Introduction

Agent-Based Simulators (ABSs) have become a popular technology for simulating the social repercussion of certain strategies. In this context, ABSs have been
developed in many domains. For instance, VoteSim [1] simulates the influences
of certain strategies on political elections. In addition, the ABS of Serrano and
Iglesias [2] simulates marketing strategies considering the spread of rumors in the
Twitter social network. Furthermore, RoboCup [3] provided a robotic ABS that
allows to make competitions between different soccer strategies. The popularity
of RoboCup has been continuously increasing up to the date. Their competitions
are international, and take place once a year.

© Springer International Publishing Switzerland 2016
R. León et al. (Eds.): MS 2016, LNBIP 254, pp. 176–185, 2016.
DOI: 10.1007/978-3-319-40506-3_18

Nevertheless, to the best of the authors' knowledge there is not any technique for designing teaching strategies based on the assistance of an ABS. In this context, the current work covers this gap by applying FTS-SOCI (an agent-based Framework for simulating Teaching Strategies with evolutions of Sociograms) [4] for designing teaching strategies and simulating their repercussion in university education.

The remaining of the paper is organized as it follows. The next section introduces the most related work to highlight the gap that the current work covers. Section 3 outlines the most relevant aspects of the current approach. Section 4 details the experiments with the current approach. Section 5 discusses some relevant aspects that have been detected in the experimentation, and indicates the main future lines of research.

2   Related Work

2.1   ABSs for Simulating Strategies

In the literature, several works use ABSs for simulating strategies. For instance, Serrano and Iglesias [2] present an open-source ABS for simulating viral marketing strategies in Twitter. In particular, it simulates rumor diffusion in social networks. In addition, they propose some strategies for controlling malicious gossips. Their approach allows companies to test certain marketing strategies before actually applying these in the real world.

In addition, Lai [5] presents an ABS that simulates strategies of buyers for preventing from sellers' cheating. In particular their agents are modeled using the mathematical model called Eavesdropping and Resistance of Negotiation Game. Their simulations show the emergent cooperative strategies between buyers and sellers.

Furthermore, DeciUrban [6] is an ABS that simulates strategies in gridding urban management. They experience different inspecting strategies in the context of Shanghai. Their simulations show the effectiveness of the strategies and the distributions of their impact. In this manner, they can design and test new urban management strategies without always needing field experiments.

In a similar way, the current approach allows teachers to simulate teaching strategies to estimate their repercussions before actually applying these with real students. However, the domain of the strategies of the current ABS is different from the ones of the previous ABSs.

2.2   Simulators in Education

Several simulators have been used for education in different subjects. To begin with, SIMBA [7] is an ABS that allows practitioners to evaluate different intelligent agents playing the role of business decision makers. Their environment is competitive and simulates common business scenarios. SIMBA has been used in business management education, so that students can better understand these environments.

Moreover, Pozo-Barajas et al. [8] present a simulator of macroeconomic models. This simulator allows students to dynamically observe their graphical evolutions, which assist them in learning and understanding these models. They compared the education scores between the students that used this simulator guided by a teacher (i.e. experimental group) and the students that learned the macroeconomic models in the mathematical traditional way (i.e. control group). They found statistically significant improvements of the students' scores when using the simulator with the guidance of a teacher.

Hence, these simulators allow students to better understand theoretical models by simulating certain cases. However, these simulators do not assist teachers in improving their teaching strategies by simulating their repercussions in the sociograms of students, as the current work does.

3   Designing Teaching Strategies in University Education

The current work is in the context of the research project mentioned in the acknowledgments section. This research project has the following steps:

1. The researchers collect sociograms in university classes related to specific training strategies.
2. FTS-SOCI is trained with the data collected in the previous step. The data is classified between training data and validation data. In this way, the simulator can be considered reliable for unknown situations.
3. The teachers can simulate their teaching strategies before applying these.
4. In fact, each teacher can consider different teaching strategy alternatives, and simulate these in order to select the one with best simulated repercussion considering some features of the resulting sociogram such as its cohesion.
5. Teachers and researchers propose possible improvements for incorporating these in FTS-SOCI.

The current work has used a survey with the three questions shown in Table 1. The goal of this survey is (a) to obtain the sociograms of each group of students considering selections and rejections, and (b) to classify students according to certain predefined types.

The current work presents the simulation of the teaching strategies of two different subjects in engineering education with FTS-SOCI. It shows both the simulated and real sociograms. For now, FTS-SOCI was not trained in the specific context of engineering education. The experimentation assisted teachers in considering alternative teaching strategies. This experimentation also allowed us to learn certain lessons, which are discussed later in Sect. 5.

FTS-SOCI was constructed following the Process for developing Efficient ABSs (PEABS) [9] with the Java programming language. FTS-SOCI provides a simple way of defining teaching strategies based on this programming language. Thus, it is usually easier to define strategies for teachers with basic knowledge about Java programming.

Table 1. Questions of the survey for detecting sociograms and classifying students

Id.	Question
Q1	Who would you choose as a workmate?
Q2	Who would you avoid as a workmate?
Q3	Which profile best defines you considering your collaborative activity in class?
	(a) quiet student,
	(b) participant student,
	(c) tangent student, i.e. sometimes you discuss topics that are tangent to the course,
	(d) joker student, i.e. you like cheering your classmates,
	(e) obstructive, i.e. you obstruct the normal functioning of the class even if you did not do it on purpose, or
	(f) occasionally participant student, i.e. you participate sometimes, while other times you act as a quite student.

In particular, the agent types are represented with object-oriented classes. Each teaching strategy is represented with a specific teacher agent type. In practice, a teaching strategy is implemented by extending the "Teacher" abstract class of the framework and implementing its "live" method. This method can call to the protected methods of the Teacher class to simulate the different teaching activities within the strategy in specific class hours.

4   Experimentation

The current approach has been experienced in two different subjects in two different engineering grades. The two subsections of this section introduce these experiences respectively for the Programming subject and the Electrical Engineering subject.

4.1   Experience with the Programming Subject

In the Programming subject, the sociogram was extracted by means of the survey presented in the previous section in Table 1. It is worth mentioning that two students replied that they would select some particular students, and that they would avoid literally the "remaining ones" without enumerating them. This response does not probably mean that they have rejection feelings towards each of the non-selected peers. They have probably misunderstood the question, and provided a general answer. Figure 1 presents the sociogram of the students ignoring this general answer.

This work has used the automatic representation of sociograms of the CLUS-SOCI tool [10] and its measurement. In this way, the real sociograms were represented with a specific text format, and CLUS-SOCI loaded these sociograms and measured their features.

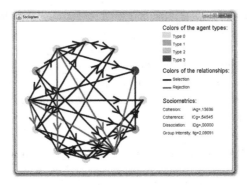

Fig. 1. Real sociogram of the students of the Programming subject (Color figure online)

```
package teacher;
/** It implements a teacher agent with the teaching strategy proposed for the
 * programming subject.
 * @author Iván García-Magariño */
public class Programming extends Teacher{
    /** This method is periodically called within the agent-based simulator,
     * and is used for defining learning strategies. */
    public void live(){
        // Variables for features of activities
        int sizeGroups; double durationHours; int interactionLevel;
        // Class exercises in certain iterations (i.e. classes)
        int classExercises[] = {20, 24, 34, 38, 46, 52, 54, 56};
        int numExercisesPerClass=10; // num exercises for each class
        for(int i=0; i<classExercises.length; i++){
            if(getIteration()==classExercises[i]){
                sizeGroups=1; // The exercises are presented individually
                durationHours=0.1; // About six minutes each exercise
                interactionLevel= Teacher.HIGH;
                makeGroups(sizeGroups);
                for (int j=0; j<numExercisesPerClass; j++){
                    roleplay(durationHours, interactionLevel);
                }
            }
        }
        // Final discussion of the whole group about the last practice
        int classDiscussion=58;
        if(getIteration()==classDiscussion){
            durationHours=2;
            interactionLevel=Teacher.HIGH;
            discussion(durationHours, interactionLevel);
        }
    }
}
```

Fig. 2. The definition of the teaching strategy for the Programming course

The FTS-SOCI was directly applied without any specific training for the engineering university context. In particular, FTS-SOCI had been training exactly as described in our previous work [4].

A new teaching strategy was defined for reflecting the strategy that was actually applied in the Programming subject. In particular, this was defined following the guidelines of the FTS-SOCI framework. Figure 2 shows the programming code that defines this strategy. It determines which collaborative activities are applied in the corresponding classes.

The FTS-SOCI was executed with the strategy for the Programming course. Figure 3 shows the main interface of FTS-SOCI for executing this strategy.

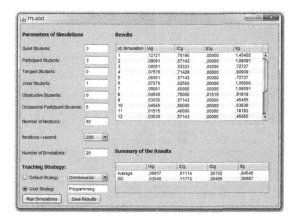

Fig. 3. The main interface of FTS-SOCI when executing the teaching strategy for the Programming course

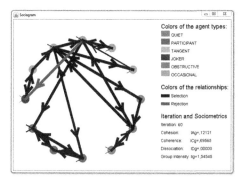

Fig. 4. An example of a simulated sociogram of FTS-SOCI when executing the teaching strategy for the Programming subject (Color figure online)

One can observe the configuration of the input parameters according to the number of students of each type and the number of classes (i.e. referred in the application as iterations). It also presents the average results of some sociometrics. Figure 4 shows an example of a sociogram simulated by FTS-SOCI with the strategy defined for the Programming course. The circles represent the students indicating their types with the color. The arrows represent the selection or rejection relationship according to the color.

The teacher of Programming considered alternative strategies. In particular, he thought to start doing the practical sessions in pairs. In this way, students can know better each other by at least knowing well its partner of practical exercises. The teacher tested this strategy by simulating it with FTS-SOCI. The definition of the this improved strategy is presented in Fig. 5. The Programming subject has four class hours each week. In particular, the practical classes are the third

```
package teacher;
/** It implements a teacher agent with the teaching strategy proposed for the
 * programming subject.
 * @author Iván García-Magariño */
public class ProgrammingImproved extends Teacher{
    /** This method is periodically called within the agent-based simulator,
     * and is used for defining learning strategies. */
    public void live(){
        // Variables for features of activities
        int sizeGroups; double durationHours; int interactionLevel;
        // Class exercises in some theoritical classes
        int classExercises[] = {20, 24, 36, 40, 48, 49, 52, 53};
        int numExercisesPerClass=10; // num exercises for each class
        for(int i=0; i<classExercises.length; i++){
            if(getIteration()==classExercises[i]){
                sizeGroups=1; // The exercises are presented individually
                durationHours=0.1; // About six minutes each exercise
                interactionLevel= Teacher.HIGH;
                makeGroups(sizeGroups);
                for (int j=0; j<numExercisesPerClass; j++){
                    roleplay(durationHours, interactionLevel);
                }
            }
        }
        // Teamwork in pairs in all the practical classes including homework
        int classWeek=getIteration()%4;
        if(classWeek==2||classWeek==3){
            sizeGroups=2;
            durationHours=2.5;
            interactionLevel=Teacher.HIGH;
            this.teamwork(durationHours, interactionLevel);
        }
        // Final discussion of the whole group about the last practice
        int classDiscussion=57;
        if(getIteration()==classDiscussion){
            durationHours=1;
            interactionLevel=Teacher.HIGH;
            discussion(durationHours, interactionLevel);
        }
    }
}
```

Fig. 5. Definition of an improved strategy for the Programming subject

and fourth hours of the week. Notice that each student spends about 2.5 h for each class hour, since they also work at home.

The teacher performed 20 simulations with the same group of students with the new teaching strategy by means of FTS-SOCI. Figure 6 shows the main interface of the application when performing these simulations. For example, one can observe that the average cohesion (i.e. IAg sociometric) has increased in comparison to the previous 20 simulations with the original strategy (compare Fig. 6 with Fig. 3).

In the literature, the cohesion of a group has proved to be directly related with its performance [11]. Thus, the new strategy can improve the performance of a group of students, according to the simulations. Hence, the current approach recommends the new teaching strategy over the original one. The teacher did not consider larger teams for practical sessions, since it is easier to control that all members of a team actively participate when the groups are smaller. The Programming subject belongs to the first semester of the first academic year of the Computer Science grade. These students need to be especially controlled to make sure that they get used to the university working rate.

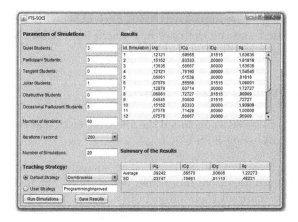

Fig. 6. Simulation results of the improved strategy for Programming subject

4.2   Experience with the Electrical Engineering Subject

The current approach has also been experienced in the Electrical Engineering subject. The teacher asked their students to fill the survey in order to obtain their sociometric status and their self-classification in the given student types.

Concretely, Fig. 7 presents the obtained sociogram of the students of this subject. Some of the students explicitly mentioned their discrepancies about replying to the question about rejection relationships. Hence, the teacher mentioned that they could optionally skip this question. Thus, the dissociation sociometric (i.e. IDg) may not be reliable since some rejection relations may be missing.

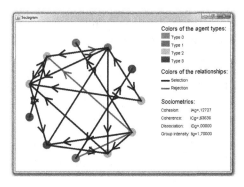

Fig. 7. Real sociogram of the students of the Electrical Engineering subject (Color figure online)

The strategy of this teacher includes three main teamwork activities in which there was only one team with all the students. Each of these activities took about eight hours, and the level of interaction was high. Besides these activities, there were other six pair teamwork activities. Each of these activities took about three hours. The first and the last ones had a high level of interaction while the remaining ones had a medium level of interaction.

5   Discussions and Future Work

The teachers and researchers concluded the following facts that can be considered in the future for improving FTS-SOCI:

- Some students share several subjects. The sociograms of students usually depend on the activities from more than one subject. For instance, first-year students usually share all their subjects in a grade. Thus, FTS-SOCI can be extended to simulate the evolution of a sociograms considering several subjects and consequently taking several teaching strategies into account. This can be accomplished by introducing several teacher agents in the simulation.
- FTS-SOCI can add more kinds of teaching activities. For instance, the framework can include a new activity type for blackboard exercises, which would be useful in engineering grades. In each of these exercises, a student presents their solution of a problem to their classmates and they can comment some aspects or ask anything.
- Some students replied to the question about rejections (i.e. question "Q2" of Table 1), literally indicating "the remaining ones" without explicitly mentioning their names. This kind of questions can alter the dissociation sociometric.
- Some teachers perceived that their students were afraid to honestly reply to the questions about the rejections. In particular, they were afraid that a classmate becomes aware of being rejected by them. They were also afraid that their teacher becomes aware of the existing rejection relationships. This fact can have influenced the veracity of the data collected about rejection relations, since students might not have felt free while answering. There are possible ways to address this problem. In the future, the surveys will be conducted by a person unknown by the students. In addition, the responses of the surveys can be collected with mobile devices for increasing the privacy of the respondents.
- The auto-classification of students sometimes is quite different from the classification from the teachers' perceptions. This was especially noticed by the teacher from the Programming subject.

As future work, the FTS-SOCI software will be officially registered and will be made publicly available, so that other teachers can use it or extend it. The authors will collect more real sociograms of students related to certain teaching strategies in different university subjects. This kind of sociograms will be also extracted from groups that are larger than the ones from the current experimentation. In this way, the current approach will be further tested by training the ABS in the same university context. The simulator will be validated with cases not belonging to the training set. Some statistical analyses will determine whether there are significant differences between the real and simulated outcomes.

Acknowledgments. This work has been supported by the project "Diseño curricular de asignaturas con actividades colaborativas asistido con un simulador basado en agentes" with reference PIIDUZ_15_193 and funded by the University of Zaragoza.

References

1. Serrano, E., Moncada, P., Garijo, M., Iglesias, C.A.: Evaluating social choice techniques into intelligent environments by agent based social simulation. Inf. Sci. **286**, 102–124 (2014)
2. Serrano, E., Iglesias, C.A.: Validating viral marketing strategies in twitter via agent-based social simulation. Expert Syst. Appl. **50**, 140–150 (2016)
3. Kitano, H., Asada, M., Kuniyoshi, Y., Noda, I., Osawa, E., Matsubara, H.: Robocup: a challenge problem for AI. AI Mag. **18**(1), 73 (1997)
4. García-Magariño, I., Plaza, I.: FTS-SOCI: an agent-based framework for simulating teaching strategies with evolutions of sociograms. Simul. Model. Pract. Theor. **57**, 161–178 (2015)
5. Lai, Y.-L.: Analyzing strategies of mobile agents on malicious cloud platform with agent-based computational economic approach. Expert Syst. Appl. **40**(7), 2615–2620 (2013)
6. Gao, L., Durnota, B., Ding, Y., Dai, H.: An agent-based simulation system for evaluating gridding urban management strategies. Knowl. Based Syst. **26**, 174–184 (2012)
7. Borrajo, F., Bueno, Y., De Pablo, I., Santos, B., Fernández, F., García, J., Sagredo, I.: Simba: a simulator for business education and research. Decis. Support Syst. **48**(3), 498–506 (2010)
8. Pozo-Barajas, R., del Pópulo Pablo-Romero, M., Caballero, R.: Evaluating a computer-based simulator program to teach the principles of macroeconomic equilibria. Comput. Educ. **69**, 71–84 (2013)
9. García-Magariño, I., Gómez-Rodríguez, A., González-Moreno, J.C., Palacios-Navarro, G.: PEABS: a process for developing efficient agent-based simulators. Eng. Appl. Artif. Intell. **46**, 104–112 (2015)
10. García-Magariño, I., Medrano, C., Lombas, A.S., Barrasa, A.: A hybrid approach with agent-based simulation and clustering for sociograms. Inf. Sci. **345**, 81–95 (2016)
11. Yang, H.-L., Tang, J.-H.: Team structure and team performance in is development: a social network perspective. Inf. Manag. **41**(3), 335–349 (2004)

Exploring the Potential of Cuckoo Search Algorithm for Classification and Function Approximation

Vasile Georgescu[(⊠)]

Department of Statistics and Informatics,
University of Craiova, Craiova, Romania
v_geo@yahoo.com

Abstract. In this paper we conduct some experiments with one of the most promising nature-inspired metaheuristic algorithm for optimization, known as Cuckoo Search (CS). It is essentially based on the cuckoo breeding behavior, which consists of dumping eggs in the nests of host birds and letting these host birds raise their chicks. The aim of this paper is to explore the performance of CS metaheuristic when it is used to evolve Neural Network for classification or function approximation.

Keywords: Cuckoo Search Algorithm · Classification · Function approximation · Evolved neural networks with cuckoo search

1 Introduction

Cuckoo search (CS) is a nature-inspired metaheuristic algorithm for optimization, developed by Yang and Deb ([5], 2009). It was inspired by cuckoos' breeding behavior and was enhanced by the so-called Lévy flight behavior associated with some birds. In the meantime, a balanced combination of a local random walk with permutation and the global explorative random walk is used, as a refined survival mechanism.

A typical cuckoos' breeding behavior refers to brood parasitism and nest takeover and may include the eviction of host eggs by recently hatched cuckoo chicks.

CS is a population-based algorithm, in a way similar to GA and PSO, but it uses some sort of elitism and/or selection. Like other population-based algorithms, CS use reproduction operators to explore the search space. Each individual (i.e., egg) represents a solution to the problem under consideration. If the cuckoo egg mimics very well the host's, then it has the chance to survive and be part of the next generation. Exploring new and potentially better solutions is the main objective of the algorithm. The randomization in CS is more efficient as the step length is heavy-tailed, and any large step is possible. Another important characteristic of this heuristic is that it dependents only on a relatively small number of parameters. Actually, the number of parameters in CS to be tuned is fewer than in GA and PSO.

Recent experiments suggest that CS has the potential of outperforming PSO and GA in terms of predictive power. Moreover, given that each nest can represent a set of solutions, CS can be also extended to the type of meta-population algorithms.

© Springer International Publishing Switzerland 2016
R. León et al. (Eds.): MS 2016, LNBIP 254, pp. 186–194, 2016.
DOI: 10.1007/978-3-319-40506-3_19

Since animals search for food in a random or quasi-random manner, their foraging path is effectively a random walk: the next move is based on the current location or state and the transition probability to the next location. The flight behavior of some birds or fruit flies has demonstrated the typical characteristics of Lévy flights, which are a form of flight that manifest power law-like characteristics. In this case, the landscape is typically explored by using a series of straight flight paths punctuated by sudden turns. Such behavior has been applied for optimization and optimal search with promising results.

2 The Cuckoo Search Algorithm

The CS heuristic can be summarized in three idealized rules:

- Each cuckoo lays one egg at a time and dumps it in a randomly chosen nest.
- The best nests with high-quality eggs will be carried over to the next generations.
- The number of available host nests is fixed, and the egg laid by a cuckoo is discovered by the host bird with a probability $p_a \in (0, 1)$. In this case, the host bird can either get rid of the egg or simply abandon the nest and build a completely new nest.

When generating new solution $x_i^{(t+1)}$ for, say, a cuckoo i, a Lévy flight is performed as

$$x_i^{(t+1)} = x_i^t + \alpha \oplus \text{Lévy} (\lambda) \tag{1}$$

where $\alpha > 0$ is the step size which should be related to the scales of the problem of interests. In most cases, $\alpha = 1$ is used. This equation is stochastic equation for random walk. In general, a random walk is a Markov chain whose next location depends only on the current location and the transition probability. The product $\oplus$ means entry-wise multiplications. The Lévy flight essentially provides a random walk while the random step length is drawn from a Lévy distribution

$$\text{Lévy} \sim \mu = t^{-\lambda}, \ (1 < \lambda \le 3), \tag{2}$$

which has an infinite variance with an infinite mean. Here the steps essentially form a random walk process with a power law step length distribution with a heavy tail. The algorithm can also be extended to more complicated cases where each nest contains multiple eggs (a set of solutions). The algorithm can be summarized as in the following pseudo code:

1. begin
2.   The objective function $f(x)$, $x = (x_1, ..., x_d)'$;
3.   Generate an initial population of n host nests (solution vectors), namely x_i $(i = 1, 2, ..., n)$;
4.   while ($t <$ Max iterations) and (termination condition not achieved)

5.   Generate a new solution vector x_{new} via Lévy flight and evaluate its fitness, say F_{new};

6.   Randomly select a vector (say, x_j) from the current population and compare the function values $f(x_j)$ and $f(x_{new})$;

7.   if $(f(x_{new}) < f(x_j))$,

8.   replace x_j by x_{new};

9.   end if

10.   A fraction (p_a) of the worse nests are abandoned and new nests are generated;

11.   Keep the best solutions (or nests with quality solutions);

12.   Rank the solutions and find the current best solution vector;

13.   end while

14.   Post process results and visualization.

15. end

- **Mantegna's algorithm**

Mantegna's algorithm [1] produces random numbers according to a symmetric Lévy stable distribution. It was developed by R. Mantegna. The algorithm needs the distribution parameters $\alpha \in [0.3, 1.99], c > 0$, and the number of iterations, n. It also requires the number of points to be generated. When not specified, it generates only one point. If an input parameter will be outside the range, an error message will be displayed and the output contains an array of NaNs (Not a Number). The algorithm is described in following steps:

$$v = \frac{x}{|y|^{1/\alpha}},$$

(3)

where x and y are normally distributed stochastic variables and

$$\sigma_x(\alpha) = \left(\frac{\Gamma(\alpha + 1) \sin\left(\frac{\pi\alpha}{2}\right)}{\Gamma\left(\frac{\alpha+1}{2}\right) \alpha 2^{(\alpha-1)/2}} \right)^{\frac{1}{\alpha}}, \sigma_y = 1.$$

(4)

The resulting distribution has the same behavior as a Lévy distribution for large values of the random variable ($|v| \gg 0$). Using the nonlinear transformation

$$w = \left((K(\alpha) - 1)e^{-|v|/C(\alpha)} + 1 \right) v,$$

(5)

the sum $z_{cn} = \frac{1}{n^{1/\alpha}} \sum_1^n w_k$ quickly converges to a Lévy stable distribution. The convergence is assured by the central limit theorem. The value of $K(\alpha)$ can be obtained as

$$K(\alpha) = \frac{\alpha\Gamma\left(\frac{\alpha+1}{2\alpha}\right)}{\Gamma\left(\frac{1}{\alpha}\right)} \left(\frac{\alpha\Gamma\left(\frac{\alpha+1}{2}\right)}{\Gamma(\alpha+1)\sin\left(\frac{\pi\alpha}{2}\right)}\right)^{\frac{1}{\alpha}}. \tag{6}$$

Also, $C(\alpha)$ is the result of a polynomial fit to the values obtained by resolving the following integral equation:

$$\frac{1}{\pi\sigma_x}\int_0^\alpha q^{1/\alpha}\exp\left(-\frac{q}{2} - \frac{q^{2/\alpha}C(\alpha)}{2\sigma_x^2}\right) dq =$$

$$\frac{1}{\pi}\int_0^\alpha \cos\left(\left(\frac{K(\alpha)-1}{e}+1\right)C(\alpha)\right)\exp(-q^\alpha)dq \tag{7}$$

The required random variable is given by $z = C^{1/\alpha}z_{cn}$.

- **Simplified version of the algorithm**

Mantegna's algorithm uses two normally distributed stochastic random variables to generate a third random variable which has the same behavior as a Lévy distribution for large values of the random variable. Further it applies a nonlinear transformation to let it quickly converge to a Lévy stable distribution. However, the difference between the Mantegna's algorithm and its simplified version used by Xin-She Yang and Saush Deb ([5]) as a part of cuckoo search algorithm is that the simplified version does not apply the aforesaid nonlinear transformation to generate Lévy flights. It uses the entry-wise multiplication of the random number so generated and distance between the current solution and the best solution obtained so far (which look similar to the Global best term in PSO) as a transition probability to move from the current location to the next location to generate a Markov chain of solution vectors. However, PSO also uses the concept of Local best. Implementation of the algorithm is very efficient with the use of Matlab's vector capability, which significantly reduces the running time. The algorithm starts with taking one by one solution from the initial population and then replacing it by a new vector generated using the steps described below:

$$stepsize = 0.01 \cdot v \cdot (s - current\ best)$$
$$new_{soln} = old_{soln} + stepsize \cdot z$$

where v is the same as in Mantegna's algorithm above with σ_x calculated for $\alpha = 3/2$. z is again a normally distributed stochastic variable.

3   Using Cuckoo Search to Evolve Neural Networks for Function Approximation

The success of NN mostly depends on their design, the training algorithm used, and the choice of structures used in training. Although Back Propagation (BP) is a very popular technique for training NNs, it suffers from two major drawbacks: low convergence rate and instability. The drawbacks are caused by a risk of being trapped in a local minimum and possibility of overshooting the minimum of the error surface. Over the last years, many numerical optimization techniques have been employed to improve the efficiency of the BP algorithm. However, one limitation of this procedure, which is a gradient-descent technique, is that it requires a differentiable neuron transfer function. Also, as neural networks generate complex error surfaces with multiple local minima, the BP fall into local minima instead of a global minimum. Evolutionary computation is often used to train the weights and parameters of neural networks. In recent years, many improved learning algorithms have been proposed to overcome the weakness of gradient-based techniques. These algorithms include global search technique such as hybrid PSO-BP, artificial bee colony algorithm, evolutionary algorithms (EA), particle swarm optimization (PSO), differential evolution (DE), ant colony, and genetic algorithms (GA).

Our aim is to use Cuckoo Search (CS) algorithm to evolve neural networks for classification and function approximation (regression), as an alternative to BP and a way to overcome the weaknesses of the conventional BP.

CS is a powerful, gradient-free, optimization algorithm that was inspired by the breeding behavior of cuckoos. It is a fast optimization algorithm in terms of number of objective function evaluations required to reach a solution. Based on some refinements or modifications introduced in the last years, it is also able to handle functions which are non-smooth, multimodal and have many dimensions. In particular, CS shows a high convergence rate to the true global minimum even at high numbers of dimensions.

3.1 Predicting the Mackey-Glass Chaotic Time Series

We choose a hard prediction problem, the Mackey-Glass chaotic time series, which is well-known for its strong non-linearity. Figure 1 shows a piece of the chaotic time series, from which we extract a training set composed by 500 samples of the Mackey-Glass chaotic time series.

The Mackey-Glass attractor is a non-linear chaotic system described by the following equation:

$$\frac{dz(t)}{dt} = -bz(t) + a\frac{z(t-\tau)}{1+z(t-\tau)^{10}}$$

where the constants are set to $a = 0.2, b = 0.1$ and $\tau = 17$. The series is resampled with period 1 according to standard practice. The inputs are formed by $L = 4$ samples spaced 6 periods from each other

$$x_t = [z(t-18),\ z(t-12),\ z(t-6),\ z(t)]$$

and the targets are chosen to be

$$y_t = [z(t+6)]$$

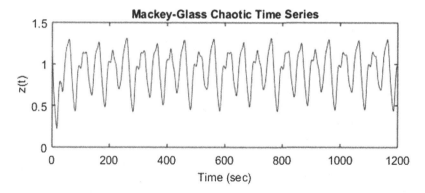

Fig. 1. Mackey-Glass chaotic time series

to perform six steps ahead prediction. Now we want to build a neural network for function approximation that can predict $y_t = [z(t+6)]$ from the past values of this time series, that is, $x_t = [z(t-18), z(t-12), z(t-6), z(t)]$.

The architecture of the neural network consists of 4 input nodes, 10 hidden nodes and 1 output node.

Figure 2 shows the observed and the predicted time series. Figure 3 shows the 3D projection of the 4-dimensional points representing the set of quality solutions in the last stage of the optimization (up) and the training performance (down).

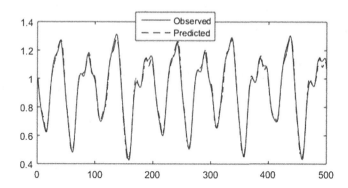

Fig. 2. Observed and predicted time series

The results in Sects. 3.1 and 3.2 have been obtained using a predefined set of internal parameters for the optimization metaheuristic. The performance can, however, be consistently improved by tuning the parameters of the Cuckoo Search algorithm. To do this, it is possible to use a secondary metaheuristic (often called meta-evolution, or hyper-heuristic) as a meta-optimization procedure, in view of finding good performing behavior parameters for the primary metaheuristic. This parameter tuning stage is attempted to improve the ability of the primary metaheuristic to approach the global optimum when training the network.

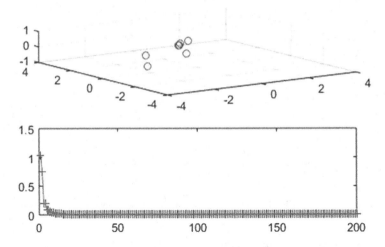

Fig. 3. Up: 3D projection of the 4-dimensional points representing the set of quality solutions; Down: training performance

3.2 Learning the Representation of a 2D Manifold

The next application consists of learning the representation of a 2D manifold (actually, a surface in the 3-dimensional space), by training a neural network using the Cuckoo Search algorithm.

Let the 2D manifold be represented by the equation:

$$z(x, y) = x \cdot e^{-x^2 - y^2}$$

We first generate a 3-dimensional point set that satisfy the above equation.

The architecture of the neural network consists of 2 input nodes, 10 hidden nodes and 1 output node.

Figure 4 shows the original 2D manifold (left) and its reconstruction (right).

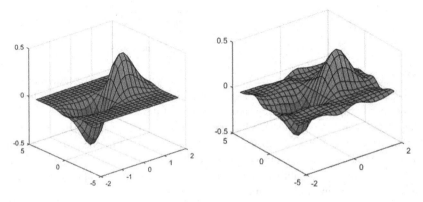

Fig. 4. The original 2D manifold (left) and its reconstruction (right).

4 Using Cuckoo Search to Evolve Neural Networks for Classification

The classification task consists of building classification models from a sample of labeled examples and is based on the search for an optimal decision rule which best discriminates between the groups in the sample.

Our purpose is to evolve neural network classifiers for bankruptcy prediction, which is a hard classification problem, as data are high-dimensional, non-Gaussian, and exceptions are common.

A sample of 130 Romanian companies has been drawn from those listed on Bucharest Stock Exchange (BSE), with the additional restriction of having a turnover higher than one million EURO. Financial results for the selected companies were collected from the 2013 year-end balance sheet and profit and loss account, and were taken from the official website of BSE, www.bvb.ro. As predictors, a number of 16 financial ratios have been used in our models.

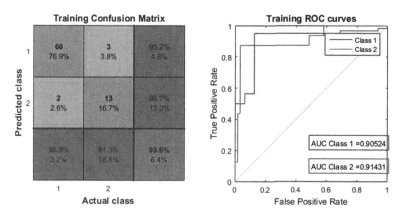

Fig. 5. NN classifier evolved with CS: training confusion matrix and ROC curves

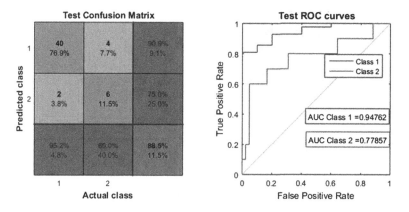

Fig. 6. NN classifier evolved with CS: test confusion matrix and ROC curves

The holdout validation method was used and the performance of the classifier was evaluated using the Confusion Matrix and the Receiver Operating Characteristic (ROC) analysis.

Figures 5 and 6 show the confusion matrices and ROC curves for training and test datasets. Given than the bankruptcy prediction is a difficult task, the accuracy of the classifier evolved with CS is very promising.

5 Conclusion

Our experiments in this paper tried to explore the potential of Cuckoo Search Algorithm for Classification and Function Approximation. The results demonstrate that CS is a powerful nature-inspired metaheuristic algorithm for optimization, which is gradient-free, has a high convergence rate and is able to handle difficult optimization problems. The algorithm performed notably well in complex task such as training the weights and parameters of neural networks for classification and function approximation.

References

1. Mantegna, R.: Fast, accurate algorithm for numerical simulation of Lévy stable stochastic processes. Phys. Rev. E **49**(5), 4677–4683 (1994)
2. Rodrigues, D., Pereira, L.A.M., Souza, A.N., Ramos, C.C., Yan, X.-S.: Binary cuckoo search: a binary cuckoo search algorithm for feature selection. In: IEEE International Symposium on Circuits and Systems (ISCAS), pp. 465–468 (2013)
3. Valian, E., Mohanna, S., Tavakoli, S.: Improved cuckoo search algorithm for feedforward neural network training. Int. J. Artif. Intell. Appl. (IJAIA) **2**(3), 36–43 (2011)
4. Walton, S., Hassan, O., Morgan, K., Brown, M.R.: Modified cuckoo search: a new gradient free optimisation algorithm. Chaos, Solitons Fractals **44**(9), 710–718 (2011). doi:10.1016/j.chaos.2011.06.004
5. Yang, X.-S., Deb, S.: Cuckoo search via Lévy flights. In: Proceedings of World Congress on Nature and Biologically Inspired Computing (NaBIC 2009), India, USA, pp. 210–214. IEEE Publications (2009)

Heating Uniformity Evaluation in Domestic Cooking Using Inverse Modelling

Fernando Sanz-Serrano[1(✉)], Carlos Sagues[1], and Sergio Llorente[2]

[1] I3A, University of Zaragoza, Zaragoza, Spain
{fer.sanz,csagues}@unizar.es
[2] Research and Development Department, Induction Technology,
Product Division Cookers, BSH Home Appliances Group, Zaragoza, Spain
sergio.llorente@bshg.com

Abstract. The heating uniformity of a domestic cooker is decisive for good results of the food elaboration. In this work we present the use of the inverse thermal modelling for the calculation of the power distribution transferred from a domestic cooker to the pan. The most common cooking stoves (induction, electric resistance and gas), have been used. The proposed method, using a rectangular thin metallic piece and an infrared camera, improves the accuracy and reduces the computational cost with respect to previous works. Moreover, it can help to improve the design process of cookers, providing more reliable results than current methods based in experimental tests with real food.

Keywords: Thermal modelling · Inverse modelling · Power distribution · Domestic cookers

1 Introduction

Manufacturers of domestic cooking appliances are continuously trying to improve the performance of their products in order to offer better services to the users at less cost. Apart from the energy consumption, efficiency and cost, one of the main characteristics which determine the good performance of the cookers is the uniformity of the temperature distribution generated in the bottom of the cooking vessel. A uniform heating ensures a good cooking result, whereas an uneven heating can originate a partial burning of the food in contact with the hottest zones of the pan. For this reason, the evaluation of the heating uniformity of each cooker during the design process becomes a necessity.

As known, in the current domestic cooktops different technologies are used as heating elements, essentially gas burners, electric resistors, or inductors. With each technology the heat is transferred to the cooking vessel by a different phenomenon, which determines the temperature uniformity and the cooking result. Current producers carry out the evaluation of the heating performance of such different systems by means of experimental tests, cooking real food [1]. These tests are costly and not easily reproducible, since the results are sensitive to the ingredients and the food composition. Thus, determination of the heating uniformity by means of other methods is of high interest. A direct method is the theoretical calculation of the power distribution

© Springer International Publishing Switzerland 2016
R. León et al. (Eds.): MS 2016, LNBIP 254, pp. 195–204, 2016.
DOI: 10.1007/978-3-319-40506-3_20

generated by each cooker depending on the technology and configuration (geometry, power supply, etc.). For example, in the case of gas burners the heat is transferred to the cooking vessel by convection and radiation from the flame and the hot gases resultant from the combustion. By means of complex models including chemical reactions and computational fluid dynamics (CFD) a heat transfer distribution in the pan base can be estimated [2]. This distribution is highly dependent on the velocity of the gas flowing through the burner outlet, which increases the complexity of the problem. In the case of electric stoves with resistors, the heat is transferred to the pan through the contact with the ceramic glass, which is heated by radiation from the hot resistance. The heat conduction problem is easier to model; however, the solution of the radiative heat transfer is typically complex to obtain due to the strong dependence with the geometry of the system and the nonlinearity of the problem [3]. Finally, the induction heating used in modern electric cooktops has been studied in several works [4–8]. Most of them have carried out numerical calculations of the coupled electromagnetic-thermal problem using finite element methods (FEM), which require high computational cost.

In order to establish a method which allows the comparison of the heating uniformity of the cookers, regardless the type of heating element, an inverse model can be used [9]. The inverse modelling has been widely used in heat transfer applications [10–12]. The main advantage which offers is the possibility of calculating non-measurable parameters such as the heat flux, heat transfer coefficients or temperatures in inaccessible locations, from the measurement of the temperature in determined positions. Examples applied to gas burners have been presented in [13, 14], in which the steady-state power distribution of the flame on the heated material is obtained with an inverse method based in analytical expressions of the temperature evolution. The applied inverse method is based in iterative calculations of the heat flux using a least squares fitting method.

In the present work, a numerical method based in finite differences is used to solve the inverse problem and compute the power distribution. This method allows the calculation of the power distribution during the transient and the steady-state modes, with different configurations of inductors, and also for other common technologies, such as gas burners and electric stoves. A heated piece with rectangular geometry is considered, and a Cartesian coordinate system is used to develop the finite differences method. Thus, the accuracy of the calculated power distribution is increased and the numerical problem is easier to implement and solve. Experiments using a gas cooker, an electric stove with a resistor and an induction hob are carried out, and the power distribution which heats the cooking vessel is calculated with the inverse model. The obtained results can be used to establish comparisons between the heating performance and to evaluate the suitability of each cooker to guarantee good cooking results.

2   Inverse Thermal Model

In [9] we presented the inverse modelling of pan heating in domestic cooktops, considering most common pans with round shape. The model was numerically solved using a finite differences method in polar coordinates. The temperature distribution on the top surface of the bottom of the pan was measured using an infrared camera, and the calculated power distribution of different cookers was presented. One of the main problems found in the calculation of the inverse problem is the strong influence of the measurement error over the results. Though it can be mitigated using image filtering, the coordinate transformation of the original rectangular thermal images into polar coordinates also introduces a noise component which produces higher deviations in the results.

The most common cooking vessels in the world's market have a round shape. Therefore, the geometry of most cookers is usually circular in order to better adapting to the pans. However, in this work we consider that the shape of the heated piece is rectangular, and the size is larger than the cooker. Hence, the geometry of the piece has negligible influence on the calculated power distribution and it is independent of the shape of the cooker. The heated piece is modelled as a thin sheet of homogeneous ferromagnetic material with known thermal properties and geometry, Fig. 1. The length in x direction is L_x, in y direction is L_y and the thickness is e.

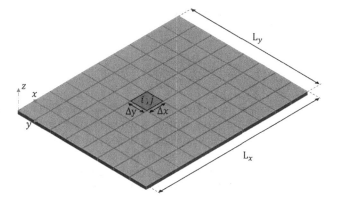

Fig. 1. Rectangular geometry of the heated piece considered in the inverse model. Length and width are L_x, L_y, thickness is e. The rectangular grid employed in the finite differences method is shown, where the size of the element i,j is $\Delta x, \Delta y$.

The temperature evolution on each point of the heated piece, placed over a heat source, obeys the well-known heat equation:

$$P + \lambda \left(\frac{\partial^2 T}{\partial x^2} + \frac{\partial^2 T}{\partial y^2} + \frac{\partial^2 T}{\partial z^2} \right) = \rho C_p \frac{\partial T}{\partial t} \tag{1}$$

where P is the equivalent volumetric power density generated by the heating source and transferred to the piece in W/m^3. The second term is the heat flux of conduction, where the Laplacian of the temperature (T) is expressed in Cartesian coordinates (x, y, z) and λ is the heat conductivity of the material. The term on the right side is the variation of the energy stored in the system with time, where the material has density ρ and specific heat capacity C_p. The inverse modelling in this thermal problem consists in obtaining the value of the power density distribution in each point of the domain and each instant, from the temperature distribution on the surface, measured during a heating process (for example with an infrared camera).

Due to the small thickness of the piece, the temperature inside the material in the z direction is considered constant. On the top surface of the plate we consider that the heat flux in z direction is equal to the losses to the environment due to convection and radiation

$$-\lambda\frac{\partial T}{\partial z} = h(T - T_a)\Big|_{z=e},$$

$$h = h_{conv} + h_{rad} = h_{conv} + \sigma\varepsilon(T^2 + T_a^2)(T + T_a) \tag{2}$$

where T_a is the ambient temperature, h_{conv} is the convection coefficient on the plate, $\sigma = 5.67 \cdot 10^{-8}$ W/m^{2}K^4 is the Stefan Boltzmann constant and $\varepsilon = 1$ is the surface emissivity, considering that the radiant surface behaves as a black body and that the surrounding surfaces are at ambient temperature. The heat flux through the bottom surface is included in the equivalent power generation term P in the case of a gas burner and the radiative electric stove

$$-\lambda\frac{\partial T}{\partial z}\Big|_{z=0} = P \cdot e. \tag{3}$$

In the case of an induction hob the bottom surface is considered adiabatic, since the gap between the coil and the heated piece will be filled with an insulating material. The power generation term embodies the heat generated from the dissipation of the induced eddy currents and the losses due to magnetic hysteresis.

The boundary conditions in the borders of the heated piece are Neumann conditions, which specify the heat flux considering a heat losses coefficient h_b including convection and radiation

$$\lambda\frac{\partial T}{\partial x}\Big|_{x=0} = h_b(T - T_a)\big|_{x=0}, \qquad \lambda\frac{\partial T}{\partial y}\Big|_{y=0} = h_b(T - T_a)\big|_{y=0},$$

$$-\lambda\frac{\partial T}{\partial x}\Big|_{x=L_x} = h_b(T - T_a)\big|_{x=L_x}, \qquad -\lambda\frac{\partial T}{\partial y}\Big|_{y=L_y} = h_b(T - T_a)\big|_{y=L_y}. \tag{4}$$

In order to calculate the power distribution, the problem is discretized applying the finite differences method, using the rectangular grid shown in Fig. 1. The sizes of each element i,j are $\Delta x = L_x/n$, $\Delta y = L_y/m$, where n, m are the number of elements in the grid in the x, y directions, respectively. A central differences scheme is used for the discretization of the spatial derivatives and a regressive approach is used to discretize the time, obtaining an implicit differences method which ensures stability and convergence. The size of the time step selected in the discretization of the temporal dimension is Δt. The power distribution in each element $P_{i,j}$ for each instant is related with the temperature measured on each element $T_{i,j}$ with the expression

$$P_{i,j} = a_{i,j}T_{i,j} + b_i^+ T_{i+1,j} + b_i^- T_{i-1,j} + c_j^+ T_{i,j+1} + c_j^- T_{i,j-1} - \frac{\rho C_p}{\Delta t}T_{i,j}^\tau - a_{i,j}^* T_a \quad (5)$$

where $T_{i,j}^\tau$ is the temperature of the element i,j in the previous instant. The coefficients $a_{i,j}, a_{i,j}^*, b_i^+, b_i^-, c_j^+, c_j^-$, take different values in the boundaries and inside the grid, see Table 1. The effect of the radiation increases nonlinearly with the temperature of the material, as seen in (2). Thus, the value of the losses factor is recalculated on each iteration, using the temperature values from the previous instant

$$h = h_{conv} + h_{rad} = h_{conv} + \sigma\varepsilon\left(T_p^{\tau 2} + T_a^2\right)\left(T_p^\tau + T_a\right) \quad (6)$$

The values of the losses coefficient h_{conv} and h_b will be estimated using measured data from experiments. Knowing the value of these coefficients and from the measured temperature, the power density supplied to each element can be obtained.

Table 1. Coefficients in the equation of the inverse model.

Index	$a_{i,j}$	$a_{i,j}^*$	b_i^+	b_i^-	c_j^+	c_j^-
$1<i<n$ $1<j<m$	$\frac{2\lambda}{\Delta x^2} + \frac{2\lambda}{\Delta y^2} + \frac{h}{e} + \frac{\rho c_p}{\Delta t}$	$\frac{h}{e}$	$-\frac{\lambda}{\Delta x^2}$	$-\frac{\lambda}{\Delta x^2}$	$-\frac{\lambda}{\Delta y^2}$	$-\frac{\lambda}{\Delta y^2}$
$i=1$ $1<j<m$	$\frac{2\lambda}{\Delta x^2} + \frac{2\lambda}{\Delta y^2} + \frac{h}{e} + \frac{\rho c_p}{\Delta t} + \frac{2h_b}{\Delta x}$	$\frac{h}{e} + \frac{2h_b}{\Delta x}$	$-\frac{\lambda}{\Delta x^2}$	0	$-\frac{\lambda}{\Delta y^2}$	$-\frac{\lambda}{\Delta y^2}$
$i=n$ $1<j<m$	$\frac{2\lambda}{\Delta x^2} + \frac{2\lambda}{\Delta y^2} + \frac{h}{e} + \frac{\rho c_p}{\Delta t} + \frac{2h_b}{\Delta x}$	$\frac{h}{e} + \frac{2h_b}{\Delta x}$	0	$-\frac{\lambda}{\Delta x^2}$	$-\frac{\lambda}{\Delta y^2}$	$-\frac{\lambda}{\Delta y^2}$
$i=1$ $j=1$	$\frac{2\lambda}{\Delta x^2} + \frac{2\lambda}{\Delta y^2} + \frac{h}{e} + \frac{\rho c_p}{\Delta t} + \frac{2h_b}{\Delta x} + \frac{2h_b}{\Delta y}$	$\frac{h}{e} + \frac{2h_b}{\Delta x} + \frac{2h_b}{\Delta y}$	$-\frac{\lambda}{\Delta x^2}$	0	$-\frac{\lambda}{\Delta y^2}$	0
$i=n$ $j=1$	$\frac{2\lambda}{\Delta x^2} + \frac{2\lambda}{\Delta y^2} + \frac{h}{e} + \frac{\rho c_p}{\Delta t} + \frac{2h_b}{\Delta x} + \frac{2h_b}{\Delta y}$	$\frac{h}{e} + \frac{2h_b}{\Delta x} + \frac{2h_b}{\Delta y}$	0	$-\frac{\lambda}{\Delta x^2}$	$-\frac{\lambda}{\Delta y^2}$	0
$1<i<n$ $j=1$	$\frac{2\lambda}{\Delta x^2} + \frac{2\lambda}{\Delta y^2} + \frac{h}{e} + \frac{\rho c_p}{\Delta t} + \frac{2h_b}{\Delta y}$	$\frac{h}{e} + \frac{2h_b}{\Delta y}$	$-\frac{\lambda}{\Delta x^2}$	$-\frac{\lambda}{\Delta x^2}$	$-\frac{\lambda}{\Delta y^2}$	0
$1<i<n$ $j=m$	$\frac{2\lambda}{\Delta x^2} + \frac{2\lambda}{\Delta y^2} + \frac{h}{e} + \frac{\rho c_p}{\Delta t} + \frac{2h_b}{\Delta y}$	$\frac{h}{e} + \frac{2h_b}{\Delta y}$	$-\frac{\lambda}{\Delta x^2}$	$-\frac{\lambda}{\Delta x^2}$	0	$-\frac{\lambda}{\Delta y^2}$
$i=1$ $j=m$	$\frac{2\lambda}{\Delta x^2} + \frac{2\lambda}{\Delta y^2} + \frac{h}{e} + \frac{\rho c_p}{\Delta t} + \frac{2h_b}{\Delta x} + \frac{2h_b}{\Delta y}$	$\frac{h}{e} + \frac{2h_b}{\Delta x} + \frac{2h_b}{\Delta y}$	$-\frac{\lambda}{\Delta x^2}$	0	0	$-\frac{\lambda}{\Delta y^2}$
$i=n$ $j=m$	$\frac{2\lambda}{\Delta x^2} + \frac{2\lambda}{\Delta y^2} + \frac{h}{e} + \frac{\rho c_p}{\Delta t} + \frac{2h_b}{\Delta x} + \frac{2h_b}{\Delta y}$	$\frac{h}{e} + \frac{2h_b}{\Delta x} + \frac{2h_b}{\Delta y}$	0	$-\frac{\lambda}{\Delta x^2}$	0	$-\frac{\lambda}{\Delta y^2}$

3 Analysis and Model Identification

The model presented in the previous section requires knowing the evolution of the temperature on the top surface of the heated piece. In order to use the model to study real cookers, experiments with real stoves have been carried out, heating ferromagnetic steel sheets and measuring the temperatures with an infrared camera. The data is used to estimate the losses coefficients on top and borders of the piece.

3.1   Experimental Set-up

Three conventional cooktops are used to perform the experiments: an induction cooker with diameter of 210 mm, an electric stove with a double resistance with diameters 123 and 213 mm, and a gas cooker with 120 mm of diameter, Fig. 2. A sheet of laminated ferromagnetic steel of 415 × 380 mm and 0.5 mm of thickness is heated with each cooktop. The thermal properties of the steel are $\lambda = 55$ W m^{-1}K^{-1}, $\rho = 7950$ kg m^{-3} and $C_p = 500$ J kg^{-1}K^{-1}. The top surface of the piece is coated with a thermo-resistant paint with emissivity $\varepsilon = 1$. The measurement of the sheet temperature has been carried out with an infrared camera (FLIR A650), which records the temperature evolution at each point with resolution of 640 × 480 pixels, at 3 fps and accuracy of $\pm$ 20°C. The camera is located at a controlled distance and oriented perpendicularly to the metallic piece in order to maximize the image resolution.

The thermographies recorded with the infrared camera in the experiments with the induction hob, the electric stove and the gas cooker, contain on each pixel the mean temperature measured in the corresponding pixel area. This information must be processed before computing the inverse model with two operations: a homography and a noise-reduction filtering. Both operations have been performed using the Image Processing toolbox of Matlab. The homography is needed to compensate the image distortion introduced by the camera lens, and it is calculated using a standard algorithm [15]. The noise-reduction filtering is necessary to improve the calculation of the temperature derivatives appearing in the inverse model, which have an amplification effect on the noise of the measured signal. In this work we propose to use a smooth filter proposed by Savitzky & Golay, [16], which is one of the most commonly used filters in image smoothing due to the improved results with respect to other filters, such as mean or median filters. The filtering is carried out in the two spatial dimensions and the filter is tuned in order to reduce the noise as much as possible.

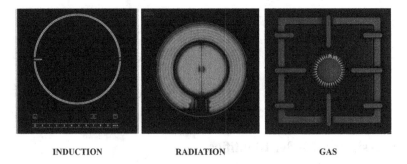

INDUCTION   RADIATION   GAS

Fig. 2. Cookers employed in the experiments: induction hob (PIE375N14E), electric stove (3EE721LS), gas stove (PPQ716B91E).

3.2   Parameter Identification

In order to simplify the identification of the losses coefficients only the induction hob is used, because the input power can be measured and controlled with higher accuracy. A power analyser (Yokogawa PZ4000) is connected in the output of the power stage which feeds the inductor and the real heating power dissipated in the sheet is estimated as the 97 % of the power measured, which is the typical efficiency of the inductor [17]. The ceramic glass which typically covers the inductor in an induction cooker is replaced with an insulating blanket of the same thickness, which is 4 mm. Thus, the bottom surface of the sheet is isolated and only the heat losses on the top surface and the borders take place.

In order to calculate the heat losses coefficient in the borders h_b, the previous expressions (4) are used, calculating the derivatives numerically from the measured temperature in the normal direction to each border during the heating process. The average value of the coefficient obtained from different experiments is $h_b = 236$ W/m^{2}K. On the other hand, the average heat loss coefficient on the top surface h_{conv} is obtained from a power balance considering the whole metallic sheet during a heating process with low power supply from the inductor, which allows for reaching a steady state at temperatures below 200°C. At the steady state the power supplied, which is measured with the power analyser, equals the heat losses. This can be formulated as

$$P_T - \sum_{\substack{1 \leq i \leq n \\ 1 \leq j \leq m}} \left(h_{conv}\left(T_{i,j} - T_a\right) + \sigma\varepsilon\left(T_{i,j}^4 - T_a^4\right) \right) \Delta x \Delta y$$

$$- \sum_{\substack{1 \leq i \leq n \\ j = 1, m}} \left(h_b\left(T_{i,j} - T_a\right)\right) e\Delta x - \sum_{\substack{i = 1, n \\ 1 \leq j \leq m}} \left(h_b\left(T_{i,j} - T_a\right)\right) e\Delta y = 0$$

(7)

where P_T is the total measured power in W. Solving the equation, the average value of the convection coefficient obtained is $h_{conv} = 9.55$ W/m^{2}K. An example of the fitting obtained is shown in Fig. 3. It can be seen how the total power losses approach to the power supplied as the steady state is reached. Moreover, the magnitude of the losses due to radiation and convection on the top surface is similar, and the losses in the borders can be neglected, as expected due to the slight thickness of the sheet.

4   Power Distribution of Domestic Cookers

The power density distributions in different heating situations with the three technologies considered in this work are calculated with the proposed inverse model. The results obtained are shown in Fig. 4, where the power distributions in the metallic piece using the induction cooker, the electric stove with a double resistance and the gas

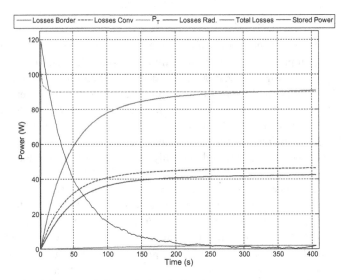

Fig. 3. Example of the power balance during the heating process up to a stationary regime using the calculated values of the heat losses coefficients. (Color figure online)

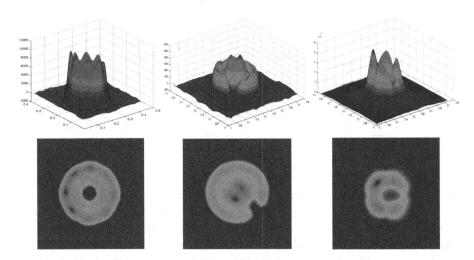

Fig. 4. Power distributions of an induction cooker (*left*), an electric stove (*middle*) and a gas cooker (*right*), calculated with the proposed inverse model.

cooker are presented. The obtained results are similar to the ones presented in [9]. However, the processing of the thermographies and the calculation is simpler and more accurate.

The calculated power distributions can help to analyse the influence of the configuration of each cooker on the heating uniformity. For example, in the case of the induction hob, the power distribution is null in the centre of the coil and has two maximums in the radial direction, which are produced by the particular distribution of wire turns of the coil. It is also observed the effect of the 8 ferrite bars beneath the coil, which amplify the magnetic field over its position, increasing the induced currents. In the double cooking zone of the electric stove, the power distribution shows a peak in the central area of the plate, originated by the higher power density radiated by the internal resistance. It can be also observed the effect of the gap between resistances, which produces a low-power area with circular shape. The power distribution transferred in the gas cooker shows a maximum where the temperature of the flame is higher, and a large central zone in which the burnt gas has a lower incidence. The metallic structure which supports the sheet above the burner isolates the bottom surface from the flame and generates gaps in the power distribution.

5   Conclusions

In this work the use of the inverse thermal modelling for the calculation of the power distribution transferred from a cooker to the pan has been presented. The most common cooking stoves, with different heating technologies (induction, electric resistance and gas), have been analysed. The proposed method, using a rectangular thin metallic piece and an infrared camera can be used to improve the accuracy and to simplify the calculation, with respect to previous works in which a polar geometry was employed. The proposed method can help to improve the design process of cookers, providing more reliable results than methods based in experimental tests with real food.

Acknowledgments. This work has been partially supported by project RTC-2014-1847-6 (RETOS-COLABORACION), from Ministerio de Economia y Competitividad, Spain, and by grant AP2013/02769 from Ministerio de Educacion, Cultura y Deporte, Spain.

References

1. AEN/CT N 213 Electrodomesticos: Household electric cooking appliances - Part 2: Hobs - Methods for measuring performance (2014)
2. Shin, J.G., Woo, J.H.: Analysis of heat transfer between the gas torch and the plate for the application of line heating. J. Manuf. Sci. Eng. **125**, 794–800 (2003)
3. Wan, X., Wang, F., Lu, Y., Huang, W., Wang, J.: A numerical analysis of the radiation distribution produced by a Radiant Protective Performance (RPP) apparatus. Int. J. Therm. Sci. **94**, 170–177 (2015)
4. Carretero, C., Acero, J., Alonso, R., Burdio, J.: Normal mode decomposition of surface power distribution in multiple coil induction heating systems. IEEE Trans. Magn. **52**, 1 (2015)
5. Sanz-Serrano, F., Sagues, C., Llorente, S.: Power distribution in coupled multiple-coil inductors for induction heating appliances. IEEE Trans. Ind. Appl. (2016)
6. Pham, H.N., Fujita, H., Ozaki, K., Uchida, N.: Estimating method of heat distribution using 3-D resistance matrix for zone-control induction heating systems. IEEE Trans. Power Electron. **27**, 3374–3382 (2012)
7. Kranjc, M., Zupanic, A., Miklavcic, D., Jarm, T.: Numerical analysis and thermographic investigation of induction heating. Int. J. Heat Mass Transf. **53**, 3585–3591 (2010)

8. Kurose, H., Miyagi, D., Takahashi, N., Uchida, N., Kawanaka, K.: 3-D eddy current analysis of induction heating apparatus considering heat emission, heat conduction, and temperature dependence of magnetic characteristics. IEEE Trans. Magn. **45**, 1847–1850 (2009)

9. Sanz-Serrano, F., Sagues, C., Llorente, S.: Inverse modeling of pan heating in domestic cookers. Appl. Therm. Eng. **92**, 137–148 (2016)

10. Plotkowski, A., Krane, M.: The use of inverse heat conduction models for estimation of transient surface heat flux in electroslag remelting. J. Heat Transf. **137**, 031301–031301-9 (2015)

11. Rodríguez, F.L., De Paulo Nicolau, V.: Inverse heat transfer approach for IR image reconstruction: application to thermal non-destructive evaluation. Appl. Therm. Eng. **33–34**, 109–118 (2012)

12. Aizaz, A., McMasters, R.L.: Detection of hot spot through inverse thermal analysis in superconducting RF cavities. Int. J. Heat Mass Transf. **48**, 4562–4568 (2005)

13. Hindasageri, V., Vedula, R.P., Prabhu, S.V.: A novel concept to estimate the steady state heat flux from impinging premixed flame jets in an enclosure by numerical IHCP technique. Int. J. Heat Mass Transf. **79**, 342–352 (2014)

14. Hindasageri, V., Kuntikana, P., Vedula, R.P., Prabhu, S.V.: An experimental and numerical investigation of heat transfer distribution of perforated plate burner flames impinging on a flat plate. Int. J. Therm. Sci. **94**, 156–169 (2015)

15. Hornberg, A.: Handbook of Machine Vision (2007)

16. Savitzky, A., Golay, M.J.E.: Smoothing and differentiation of data by simplified least squares procedures. Anal. Chem. **36**, 1627–1639 (1964)

17. Acero, J., Burdio, J.M., Barragan, L.A., Navarro, D., Alonso, R., García, J.R., Monterde, F., Hernandez, P., Llorente, S., Garde, I.: Domestic induction appliances. IEEE Trans. Ind. Appl. Mag. **16**, 39–47 (2010)

Author Index

Printed in the United States
By Bookmasters